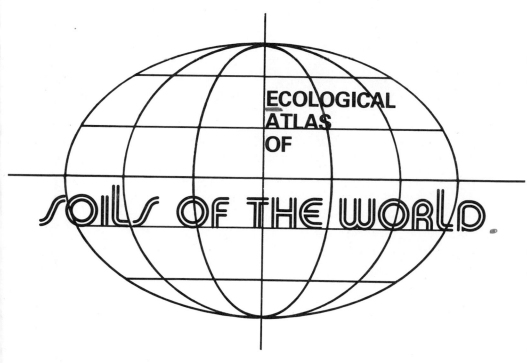

ECOLOGICAL ATLAS OF

ʃOILʃ OF THE WORLD

PHILIPPE DUCHAUFOUR

Director, Centre de Paedologie, C.N.R.S.
Professor, University of Nancy I, France

Translated from the French by

G.R. MEHUYS, C.R. DE KIMPE, and Y.A. MARTEL
Agriculture Canada Research Station
Ste-Foy, Quebec, Canada

MASSON Publishing USA, Inc.
New York • Paris • Barcelona • Milan • Mexico City • Rio de Janeiro

This volume was published with the assistance of the Centre National de la Recherche Scientifique.

Original French Edition:
ATLAS ÉCOLOGIQUE DES SOLS DU MONDE.

Other Translations:
ATLAS ECOLÓGICO DE LOS SUELOS DEL MUNDO
published by Toray-Masson, Barcelona.

ISBN 0-89352-012-8

Library of Congress Catalog Number: 77-94822

Printed in the United States of America

Foreword

In welcoming and commending Philippe Duchaufour's epoch-making volume to the English-speaking world, it is necessary to understand a little of the historical background. During the last two centuries, the great disciplines of the natural sciences have all had to face kindred problems relating to classification, terminology, and nomenclature. In their different ways, every science that dealt with either the living creatures or the physical nature of the earth was forced to accept the need for a field methodology that involved recognition, description, and cataloging.

The first and foremost classifier was Carl Linneus, who gave us the earliest effective system for a botanical nomenclature, and with that secure base, hand in hand with the growth of knowledge, hierarchical classifications could gradually evolve. Each of the earth science disciplines—mineralogy, petrology, geomorphology, oceanography, meteorology—has encountered different problems, but each has struggled to evolve a system of labels, then a classificatory hierarchy, and finally a method of naming units observed in the field.

Soil science has been no exception. While it was the Industrial Revolution that led to the start of mineralogical and geological mapping in France, Germany, and England, followed closely by the United States, the historical grounds for the establishment of soil science are not so evident. Clearly the foundations were laid in Czarist Russia, and owed much to the traditionally high standards in chemistry that had already evolved there in the last century. But more important perhaps is the fact that European Russia was agriculturally oriented, is physiographically flat and yet is very distinctly zoned in terms of climate and ecology.

From this beginning, an Eastern European, Russo-German school has evolved and been extended overseas, especially into Asia and Africa, with the notable addition of the French. It is not surprising that the F.A.O., with its headquarters in Rome and its predominantly Asiatic, African and Latin American responsibilities, has developed a "European" school. And this is reflected also by the work towards the legend of the International Soil Map of the World led by Dr. Viktor Kovda of the U.S.S.R.

Meanwhile in North America, under a strictly pragmatic agricultural aegis, the United States had developed an almost totally different methodology. The first had an essentially genetic orientation; the second was analytic. Workers in Britain, Australia, and elsewhere in the English-speaking world have been understandably confused by the multiple systems, and for their own needs have developed separate approaches. Australia perhaps has been most instrumental in leading the way towards a recognition of the role of geologic history in defining a soil; while large parts of the continent have been evolving as an undisturbed, physically stable land mass since its last glacial age (the Permian, around 250 million years ago), it has also been subjected to a geologically rapid transit, due to continental drift through about 25° of latitude during the last 50 million years, attended by the violent climatic fluctuations of this era.

The stage is thus set for a new approach to the question, and this is provided by Philippe Duchaufour with his *ecologic method*. He employs a dynamic–genetic philosophy, taking into account the parent material (the mineralogical base), the climatic environment, and the biological factors. In this volume he briefly sets out this theoretical base in a dozen pages or so, and then proceeds with a catalogue of standard soils. Each is provided with its F.A.O. and U.S. equivalents, a type locality with its topographic description, parent rock, climate and vegetation. There follows a descriptive profile (with soil morphology and color photo), then a geochemical/biochemical summary (granulometry, etc.), its "evolution", and finally a basic reference. Soils are grouped into ten chapters and a final Appendix deals with Structures and Microstructures. A glance at the brief but essential bibliography shows that the work, although oriented towards France, has drawn extensively on material from the U.S.S.R., Germany, the United States as well as South America, Africa, and so on.

I believe that the English-speaking soil scientists will appreciate and reap considerable benefit from this unique volume.

Rhodes W. Fairbridge
Department of Geological Sciences
Columbia University
New York, New York

Preface

The third edition of *Précis de Pédologie*, published in 1970, presented an attempt to classify soils on the basis of their ecology which was in fact quite close to the French soil taxonomy. The main features of the soil profiles and their dynamic genesis were explained, insofar as possible, by environmental factors. However, these concepts remained very theoretical as no examples were studied in detail. The main purpose of this Atlas is precisely to fill this void and to complete the *Précis* by describing "typical profiles" chosen from among the most characteristic. Each soil entry containing a description and analytical data has been written as concisely as possible and is illustrated by a color photograph of the profile.

In addition, the publication of this book has enabled me to state even more explicitly than in the *Précis* my views on the ecological method applied to the study of soils. In my opinion, soil classifications based on genetic and dynamic concepts should embody all environmental factors, be they bioclimatic or mineralogical (the latter relating more directly to the concept of "parent material"). It is in this frame of mind that I have tried to show by the use of "ecological tables" that soil profiles can be more easily inserted within chained "evolution sequences" than in rigid compartments. Far from supplanting existing soil classifications, I feel that the tables could be used as guidelines and may aid in explaining these systems to a certain degree.

With my objectives now well in mind, the reader should not be surprised by the necessary incompleteness of the Atlas. It does not aim at presenting the *totality* of existing soil profiles, as would be the case of a catalog or a "flora" for plants. In fact, a choice has been made by selecting the most demonstrative profiles as much because of their ecological and scientific significance as for their didactic value. Many were chosen to illustrate the ecological method with concrete examples. The many omissions in the book are due in large part to my insufficient knowledge of the soils of several regions of the world. I trust the reader will be forgiving and I hope to be able to fill some of the gaps in the future.

The Atlas owes much to many research establishments in France, as well as abroad. Its successful completion would have been impossible without the contributions of the scientists and technicians of the Centre de Pédologie, Centre National de la Recherche Scientifique (C.N.R.S.), Nancy. As it were, the book summarizes the work of the Centre, especially with respect to the genesis of temperate soils. I am most grateful to all staff members of the Centre—scientists, technicians, and secretaries—for their splendid cooperation.

Furthermore, a large use has been made of the work of the research teams of the Institut National de la Recherche Agronomique (I.N.R.A.), especially those collaborating on the elaboration of the Soil Map of France, and of the publications of the Office de la Recherche Scientifique et Technique Outre-Mer (O.R.S.T.O.M.) for all tropical soils. Finally, several profiles were described and analyzed by the scientific societies of many countries including Colombia, the U.S.S.R., Spain, West Germany, Romania, Austria, Italy, the U.S., Morocco, etc. I express my sincere thanks to all these organizations, French and foreign, and to the Soils Division, F.A.O. (Dr. Dudal), for all the data with which I was provided.

Finally, my gratitude goes to Professor Kovda and to the many Russian soil scientists who supplied precious information about the zonality of soils of their country.

Philippe Duchaufour

CONTENTS

COLOR PLATES

Plate

OBJECTIVE AND METHODS

PURPOSE OF THE ATLAS

It is not the purpose of this book to introduce another system of soil classification. Indeed, in presenting the different soil profiles, I have followed the general scheme of the French system of soil classification which is quite suitable, as it is, undoubtedly, one of the most "ecologically" oriented. At most, I have modified some divisions within its classes in order to make it even more closely reflect ecology.

My objective, which is quite different, is essentially to show that a soil cannot be truly defined outside the *environment* within which it has formed. In itself, a soil constitutes a complex "environment," generally in equilibrium with all the ecological factors such as climate, vegetation, and mineral parent material, whose actions it integrates. This state of equilibrium was not achieved readily, but through either fast or slow, simple or intricate evolution which was itself conditioned by all the environmental factors. *Features* taken into account to define a soil are the end products of this *ecological evolution*. Therefore, these features—be they physical, geochemical, biochemical, or biological—are coordinated and together form a harmonious unit which constitutes the *soil profile*, of which each "horizon" is related by "genetic links" to the others.

I will thus adopt Schröder's model (1973) which suggests that the trilogy environment → process → features be taken into account to construct a genetic classification. This concept is very close to Gerassimov's (1974a), which is based on the idea of "elementary pedological process," a synthesis of the physicochemical processes defined by environmental conditions.

It is soon evident, however, that the formulation of an ideal system of soil classification based on these principles is not feasible. First, many processes of ecological evolution are neither well known nor completely understood. Second, as any classification system aims to separate groups of soils into more or less artificial compartments at various levels, which imply cutoff points, it must adopt an inevitably too rigid framework which incompletely reflects the actual facts. As we shall see, there are many transition soils (so-called *intergrades*) that lie at the junction of two classes, fitting equally well in either one. Furthermore, in certain very

1

complex environments, soil formation depends on several distinct evolutionary processes. Soils thus formed would fall into different classes (inasmuch as a class is defined by a basic "ecological process" such as in the French classification system). In these two examples, a necessarily arbitrary choice must be made.

Such a drawback is inherent to any system of soil classification but more particularly to those whose aim is more *practical* (soil mapping) than scientific, i.e., systems based on soil characteristics that are defined, often with great precision, by essentially statistical methods. Such is the case in both the F.A.O. and U.S. systems of soil classification. When the statistical characteristics taken into account relate to the environment (happily, several diagnostic horizons of these two systems do), the scope of the "ecological" classification systems is met and correlations are easily established. On the contrary, any correlation becomes impossible when key features are chosen independently of the environment. Thus, I have attempted in this book to establish a correlation between my own nomenclature and that of the F.A.O. and U.S. systems. The reader will note that often the same name is used for profiles which are in fact very different based on their ecology and their type of evolution. On the other hand, dissimilar names (even at the order or suborder level) are given to soils whose ecology and genetic features appear to be very close. This means that the criteria used to designate these soils were not based on *ecological processes of evolution*.

A truly ecological system of soil classification should not be based on a "pyramidal" array of arbitrary classes (or orders) and subclasses (or suborders), but rather on "evolutionary chains," called *evolution sequences*, established in relation to the various environmental factors. Each soil can then be assigned a place in relation to others from which it differs either by an intensification *of the basic ecological process* (thus by a more or less strong degree of differentiation and evolution), or by *superimposition of another process to the basic process* which will set it into another class. This is what I have attempted to do in the various "ecological tables" throughout the book. These tables must not be considered independently. They complement each other and exemplify the progressive transition of many ecological units from one class to another. It will thus be evident to the reader that not too much importance should be given to the location of "intergrade" units in one class or another. It really matters little whether a Vertic Alluvial soil is considered to be an Alluvial soil or a Vertisol, or whether an Eutrophic Brown soil developed on terra fusca is placed in the Calcimagnesian or into the Brunified class. I consider it more important that the reader understand the *ecological genesis* of the profile from which originate its transitional features and consequently the problems it causes the pedologist.

Although the soil profiles are presented in a relatively new ecological light, the purpose of the Atlas is not to replace the various systems of soil classification that exist today while exposing their sometimes artificial characteristics. Obviously, these systems, even if they are imperfect, are indispensable because they are a necessary tool to the soil surveyor or the soil mapper. Practical applications of pedology are important and varied: in forestry, agronomy, urban planning, etc. These uses are based essentially on *soil maps* at various scales which require delineation of *mapping units* at different levels. Such mapping units are specifically provided by the classification systems. Nevertheless, existing systems of soil classification will progress only if they can be brought into closer agreement with the dynamic ecological concepts that I am trying to define. The users themselves will be able to introduce the necessary corrections. Aren't soil maps also essential to *pedological research*, providing it with a starting point, a natural setting, and an ecological support? Only with soil maps can research, by ingeniously combining field studies and laboratory analyses, reveal the ties of the trilogy environment → process → features. In my opinion, pedological research must evolve along this line, if its fundamental goal is indeed, as I believe it should be, the creation of a cogent ecological system of soil classification.

HOW THE PROFILES ARE PRESENTED

I decided to condense *each profile description to a single page*. This led me to adopt succinct descriptions and comments, an inconvenience that is largely counterbalanced by the possibility of a synthetic interpretation and a general view of the genetic features.

Each profile is designated by a Roman numeral which refers to the *plate* number and a subscript Arabic numeral referring to the *profile* in the plate.

The presentation of the profiles is divided into four parts:

1. Location and ecology.
2. Description and morphology.
3. Geochemical and biochemical properties.
4. Comments about the genesis of the profile as it relates to environmental factors.

Profiles which are studied in the text are outlined in the tables.

Location and ecology

In addition to information on the *geographical situation* of the profile, all soil entries contain useful details on the following:

General ecological conditions: climate, annual precipitation, mean annual temperature, occurrence of a dry season.

Local ecological conditions: parent rock or "parent material," unmixed or complex; topography. When appropriate, situation of the profile in a "catena," of which several profiles may be compared with respect to their position in the catena. In some cases, and if important, characteristics of a *local climate* are given in the description of the site.

As far as possible, data are also provided on the *native vegetation*, reflecting either the general climatic factors or even the local site conditions. According to Duvigneaud's scheme (1974), the former describe biomes or general plant formations, while the latter correspond to "regional associations," or even to "site associations," in which "ecological groups" of index-plants allow a sound characterization of the kinds of humus.

Profile descriptions

Descriptions are purposely simplified compared to those prepared by soil surveyors. Only the main diagnostic features are presented: texture, color (Munsell notation), structure, consistence, porosity, biological activity, root development, distinctness, and form of the "boundary" with the next horizon. When possible, data on the *microstructure* are added in the description, or sometimes even summarized in a paragraph (giving, for example, the form and the nature of the "coatings" of either oriented clay or "argillans," or amorphous materials—organic matter and sesquioxides).

Horizon designation. Particular attention was paid to horizon designation. Its principles are summarized below (Duchaufour, 1970, pp. 221-222).

Master Horizons

A: Surface horizon, containing organic matter, often depleted in fine particles or iron through eluviation.

(B):   "Structural" or alteration B, differing from the parent material by a higher degree of weathering (presence of free Fe_2O_3) and from the surface A horizon by a different structure (corresponding to B_v).

B: Horizon enriched with fine or amorphous particles through illuviation: clay, iron, and aluminum oxides, sometimes humus.

C: Parent material from which A and (B) or B horizons have formed.

G: Greenish-gray horizon, rich in ferrous iron, with mottles developing within or at the upper boundary of permanent ground water.

R: Consolidated bedrock.

Subdivision of Master Horizons

A Horizons:

A_{00} (or L): Litter with identifiable plant fragments.

A_0:   Organic horizon where the original structure is destroyed (contains more than 30% organic matter).

A_1:   Mixed horizon containing organic matter (less than 30%) and mineral particles (corresponds to A_h).

A_p: Humic plowed horizon, homogenized, with a clear lower boundary.

A_2:   Horizon low in organic matter, often depleted of clay and sesquioxides, lightly colored (so-called "eluvial" horizon) (corresponds to A_e or E).

A/B: Transition horizon between the eluvial and illuvial horizons with a beginning accumulation of fine or amorphous particles.

B Horizons:

B_t: Clay accumulation.
B_h: Humus accumulation.
B_s: Mainly sesquioxide accumulation.
B_b: Placic horizon: involute aliotic band.

NOTE: Numeric subscripts indicate variations in the aspect of the horizon or in the degree of accumulation.

Some subscripts apply to all A, B, or C master horizons.

g: Pseudogley, with temporary waterlogged conditions; mixture of gray, white, and rust mottles, sometimes black concretions.

ca: Horizon enriched in calcium carbonate.

x: Fragipan.

Special Horizons. Attention is drawn to some complex horizons which are particularly important for diagnostic purposes:

A_0H: "Humified" layer in a mor horizon, with the original plant material very strongly transformed, as opposed to A_0F, or "fermentation layer."

A_1B: Horizon with "mature" humus resulting from the insolubilization of certain organomineral complexes, which is characteristic of some humic profiles (Rankers, Andosols, Humic Ochric soils).

B_tB_s: Horizon with mixed "argillic" and "spodic" features.

Beta (β): Special horizon of alteration and accumulation, rich in fine clay and iron, developed above certain hard limestone layers and underlying loamy or sandy material.

Analytical methods: geochemistry and biochemistry

Soil analyses were carried out at the *Centre de Pédologie, C.N.R.S.* Nancy, following the methods described and discussed in the third part of *Précis de Pédologie* (Duchaufour, 1970). In the following section, references are given to detailed analytical methods which can be found in the Appendix (pp. 435-448) of that book, to which the reader is referred.

Particle-size distribution. Texture determination (Duchaufour, 1970, p. 31). This method involves a complete destruction of the organic matter (Na-hypochlorite), carbonate removal with HCl when applicable, and iron removal with the "combined" reagent (Tamm plus Na-dithionite). Determination of texture itself is based on the Mériaux densimetric method. The *translocation index* (T.I.) corresponds to the ratio of *clay* percentage in the A horizon to that in the B horizon. The same index, determined for free iron and aluminum, characterizes the extent of migration in a homogeneous material.

Exchange complex. The exchange capacity (CEC or T), the sum of exchangeable bases (S), and the base-saturation percentage (S/T) were determined in most soils by extraction with $1N$ ammonium acetate (p. 438, No. 6). Exchange capacity values, T and S, are expressed in milliequivalents (meq) per 100 g of material. In very humic soils, the method involves the use of buffered $CaCl_2$ (p. 438, No. 7).

- pH values were measured in *water* at a 1/1 soil/water ratio.
- Total carbonates were determined with the Bernard calcimeter (p. 440, No. 10).
- Active carbonate was measured following the Drouineau method (p. 441, No. 11). When active carbonates are present, the exchange complex is generally base-saturated.

Biochemistry. Study of the organic matter (O.M.) in the organomineral complexes: organic carbon was determined as CO_2 with the "carmhograph" after sample combustion. Total organic matter content is obtained by multiplying the carbon content by 2 (forest soils) or by 1.72 (cultivated soils).

Determination of total nitrogen by the Kjeldahl method (p. 436, No. 3) allows the C/N ratio to be calculated. This is a good indication of the total biological activity in the humus.

In some soils, the extent and nature of humification were studied in more detail (p. 425; p. 437, No. 5). The usual method gives the contents of fulvic (FA) and humic (HA) acids, from which the extractable carbon, defined as C (FA+HA)/total C in terms of percent, and also the FA/HA ratio characteristic of some humus are calculated. The degree of polymerization of humic acids (gray and brown HA) was investigated either by paper electrophoresis (Duchaufour and Jacquin, 1966) or by studying "molecular space-filling," using "Sephadex" molecular sieves.

The nonextractable humus fraction, or humin, was separated from the fresh and only slightly transformed organic matter by densimetry in heavy liquid, using ultrasonic vibration (p. 452, Monnier method). Scientists at the Biochemistry Section of the *Centre de Pédologie* have identified three fundamental kinds of humin: *residual* or *inherited humin*, resulting from the incomplete transformation of some plant debris; *microbial humin*, made up of polysaccharides; and *insolubilization humin*, resulting from the evolution of phenolic polymers precipitated by cations (Duchaufour, 1973).

Iron and aluminum oxides. Their importance in the "genetic" interpretation of profiles has prompted a great interest in their study, both qualitative and quantitative.

So-called "free" sesquioxides were extracted by the "combined reagent" proposed by the Biochemistry Section of the *Centre de Pédologie* (p. 442, No. 13).

"Complexed mobile" forms of iron and aluminum were determined following the Bruckert method, using a NaOH solution at pH 9.8, buffered with Na-tetraborate. This reagent is more selective than $0.1M$ pyrophosphate (Bruckert and Metche, 1972), which also extracts some clay-bound, and thus noncomplexed, free iron. The complexed mobile iron/total free iron ratio is a good index used to distinguish soils in a podzolic sequence (e.g., Ochric soils) from Acid Brown soils, where complexed forms are not very abundant in the (B) horizon (Bruckert et al., 1975).

The free to total iron ratio has been used as a weathering index in many soils. (See the comparison between temperate Brown soils and Fersialitic soils.)

The free aluminum to clay percentage ratio gives an indication of clay degradation under very acid soil conditions.

NOTE 1. *Values of iron and aluminum are expressed as elemental Fe and Al percentages (only "allophane" in Andosols and oxides from tropical soils are expressed as Fe_2O_3 and Al_2O_3 percentages).*

NOTE 2. Several analytical results were obtained using methods differing slightly from those reported above. For instance, free iron and aluminum oxides were determined in some soils according to the Deb or Jackson procedures. However, we have verified that results thus obtained were closely related to ours and permit valid comparisons. Most of the analytical results given here were indeed obtained using international methods, accepted by most laboratories in several countries.

Use of morphological and analytical data: diagnosis

The results of morphological and analytical investigations should not be considered individually, but must be related to each other in order to formulate a sound "diagnosis" of the place a soil should occupy in a classification system or preferably within an "evolution sequence" characteristic of a well-defined *environment*.

Partial interpretations are given both in the description and in the presentation of the geochemical and biochemical features of the profiles. These are necessarily incomplete. The general interpretation is a synthesis based on comparisons of all ecological, morphological, and analytical data. It is presented at the bottom of the soil entries, under the heading "Genesis."

The concept of "diagnostic horizon," taken from the F.A.O. and U.S. systems of soil classification, has been widely used. Selected diagnostic horizons with recognized "genetic" characteristics form an essential part of my interpretations. The original definitions, although slightly simplified, have been maintained (Duchaufour, 1970, pp. 174, 185). Examples are the *argillic* horizon (B_t) characteristic of temperate "Eluviated" soils, the *spodic* horizon (humic B_h or sesquioxide-rich B_s) characteristic of Podzolized soils, the *oxic* horizon of alteration in Ferralitic soils, the *natric* horizon, typical of Solonetz, etc. Some secondary diagnostic horizons have also been considered, such as the *calcic* horizon, *fragipan*, *plinthite*, *albic* horizon, etc.

However, I do not consider that one horizon only, even very precisely defined, is sufficient to formulate an overall diagnosis. The various horizons of a profile should not be considered separately. Each one will influence the formation of the other horizons during the genesis of the profile. Thus, the internal genetic bond existing between horizons becomes the basis of the diagnosis. Therefore, any ecological classification system must necessarily consider the whole profile. On the other hand, as I will demonstrate in Chapter I, the ecological definition of a soil depends largely on soil-plant equilibria. Obviously, soils which have aged for a long time under a permanent and stable vegetation (forest, for example) best reflect this state of equilibrium and therefore were chosen as soil types in preference to cultivated profiles. For the same reason, I have paid particular attention to the *kind of humus* which is known to perfectly integrate environmental and vegetation conditions. It often has a determining influence on the development of mineral horizons.

Nevertheless, it is also important to describe and classify cultivated profiles—those with an anthropic character at least in the surface horizon. I believe that *such a classification must be made by comparison with similar profiles, but aged under permanent vegetation*. Hence, the new features acquired by cultivation become additional diagnostic elements and allow for discrimination between the cultivated and the homologous, non-cultivated soil profile. For example, a distinction will be made between *resaturated Acid Brown soils* (i.e., through cultivation) and *Eutrophic Brown soils*, in which the high base-saturation level depends on the initial conditions of the environment. Similarly, I will describe an *Anthropic Brown Rendzina*, which differs from Forest Rendzinas by a large decrease in organic matter content (thus by a change in color) and by an increase in active carbonate content.

NOTE: Summary of the main abbreviations used in the text:

O.M.:	Organic matter.
S/T:	Base saturation (in %) of the exchange capacity (T or CEC) in exchangeable bases (S).
FA:	Fulvic acids.
HA:	Humic acids.
P.:	Annual precipitation.
M.T.:	Mean annual temperature.

Except where noted, photographs were taken by the author.

CHAPTER I

GENERAL: BASIS OF AN ECOLOGICAL SYSTEM OF SOIL CLASSIFICATION

CONCEPT OF SOIL CLIMAX AS THE BASIS OF AN ECOLOGICAL SYSTEM OF CLASSIFICATION

Areas where mineral materials have been recently deposited, or where rock has been exposed by erosion, are progressively invaded, first by herbaceous plants, then by shrubs, and finally by trees. A pedological "profile" slowly develops. Superficial at the beginning, it evolves by incorporation of organic matter (AC profile), then by weathering [A(B)C profile], and finally by migration and redistribution of soluble compounds or fine particles (ABC profile). After a variable length of time, a *state of equilibrium* between soil and vegetation is attained. "Climax" is the term often used to refer to a *stable vegetation* under the influence of all environmental factors and undisturbed by man. This term can also apply to the soil which is in equilibrium with the plant climax and thus to the entire "ecosystem."

A soil and vegetation "climax" contrasts with a "pioneer" or "transitory" vegetation developed on young soils and also with a soil and vegetation that have been "degraded," i.e., anthropic in nature. For example, I will draw a clear distinction between "primary" Podzols, which are climax soils in equilibrium with the original vegetation (Boreal or Subalpine Podzols), and "secondary" Podzols, which are degraded Forest Brown soils because of man's substitution of the primitive deciduous forest by an Ericaceae association.

ECOLOGY AND SOIL GENESIS

A given soil reaches equilibrium as a result of the interaction of various environmental factors. These can be separated into two fundamental groups: (i) *general bioclimatic factors*, i.e., the "general" climate of a given area and its so-called "climatic" vegetation which is relatively independent of local ecological factors, topography, parent material, etc., constituting plant formations or biomes and, on a larger scale, regional associations (Duvigneaud, 1974); and (ii) *"site" factors* which correspond to rock outcrops or are influenced by topography as long as the general climatic conditions are not modified as in the case of mountainous regions.

Impact of general bioclimatic factors

The general climate and plant formations (or biomes) that, on a world scale, characterize vegetation "zones" (or "elevation zones" in mountainous regions: see Profiles I_1 and I_2) are particularly important concerns of pedology. The pedogenesis of vast regions with a specific climate and type of vegetation is directly influenced by such bioclimatic factors and has led the Russian school to formulate the concept of soil "zonality," which can be found in several places in this book. Zonality implies a prominent action of climatic vegetation over parent material. This action determines the general orientation of soil evolution in a given climatic region. Vegetation acts on soil through the "type of humus" it promotes. The nature and amount of organic litter added to the soil affect the *humification process* and the *biogeochemical cycle* of cations. Cations are either concentrated and immobilized in the surface humic horizons or translocated down through the profile. *Organomineral complexes*—some with large molecules which are rapidly precipitated at the surface, others with small mobile molecules which tend to migrate down the profile—play a significant role in soil genesis.

Organic matter plays a considerable role as an intermediary between vegetation and mineral matter and as such acts as the "driving force" of soil genesis. Because of the great importance given recently to biological factors, the concept of "analogous soil" has been added to that of "zonal soil" (Pallmann, 1948). Analogous soils are formed from different parent materials but under the same climatic vegetation. By this definition, humus formation will tend to be comparable and will result in rather uniform upper horizons. Yet deep mineral horizons will reflect the possibly very different properties of the mineral material (Profiles I_1 and I_2).

The term "climatic climax" has been proposed to describe soil-vegetation equilibria which are governed mainly by the general climate. These are characterized by their "humus," whose integrating effect on the environment is now well-established.

Influence of site factors

However, all soils within a given climatic zone are far from being identical or even "analogous." More frequent are specific states of equilibrium dependent on the *local environmental conditions* of individual *sites* whose importance is revealed in large scale studies or maps. Site factors include mainly the nature of the parent material, local topography, and drainage conditions ("leaching" or "confined" medium). When site conditions, such as parent material or drainage conditions, vary only within limited bounds, the development of a climatic vegetation and the formation of "analogous soils" are not prevented. This is the general case. However, beyond certain physicochemical "thresholds," a specific vegetation (sometimes called a "specialized association") takes the place of the climatic vegetation and favors a different humification and thus a different soil formation process. This is a "site climax." Examples of site climaxes are numerous, e.g., marshland and peatland which result from poor drainage. Very specific soil climate conditions can also play an important role (Profile I_3). Some profiles reflect a particular composition of the parent material, such as "Rendzinas" which have developed, relatively independently of climate, from a material high in active carbonates. Finally, only very rarely do site climaxes originate exclusively from biological action. For instance, the particularly acidifying litter of a *single tree*, which has stood for several centuries, may create a Podzol in the shape of an "egg-cup" or "funnel" (Profile I_4).

Many soils at equilibrium are of "mixed" origin, and their formation requires the simultaneous presence of certain general bioclimatic conditions, as well as local site conditions. Such is the case of *Vertisols* and *Planosols*, which are found only in areas with a contrasting climate (with a very dry season), temporary waterlogged conditions, and parent material composition found only at very definite "sites."

Concept of parent material: mono- and polycyclic soils

The concept of "parent rock" has recently been completed and modified by new developments in Quaternary geology. More and more, the concept of "original material" is taking its place. No problem exists with soils developed from recently deposited (or eroded) mineral materials, such as postglacial material in temperate climate soils. The initial "weathering" phase and the bioclimatic genesis of the soil (particularly humus formation) occur almost simultaneously. Such a soil, generally expressing well an equilibrium with the present climatic vegetation, is called a "monocyclic" (or monogenetic) soil. The term "primary" soil is also frequently used, although it conveys an improper and confusing meaning.

Frequently, however, there exists, between the unweathered rock of old formations and the soil *per se*, an intermediate "parent material" which may have its own long and complicated history. Often, this material has undergone earlier weathering unrelated to the present climatic and vegetation conditions. Such are *Paleosols* whose features have been *inherited*, at least in part, from ancient soil forming processes (Profiles I_5 and I_6). These processes have often been interrupted by mechanical reworking of the profile by such processes as erosion, cryoturbation, etc., rendering soil interpretations yet more complicated. The present development is "superimposed" on a former, but different, evolution. Such soils are called "polycyclic" (or polygenetic) soils. Less accurately, the term "secondary soil" is often used, but it must be remembered that this also refers to biologically "degraded soils."

In such a soil, current features resulting from the prevailing bioclimatic factors are added to "inherited" features (which are usually related to the nature of the weathering products). Examples of current features are the kind of humus, the migration processes of soluble organomineral complexes, etc. Profiles I_2 and II_1 illustrate particularly well such successive evolutionary phases.

Polycyclic soils in which successive soil-forming processes are exclusively biological are rarer because humus develops rapidly and seldom shows stability. Nevertheless, Profiles II_2 and II_3 are two examples where successive soil-forming processes have occurred as a result of different humification phases.

Cases where successive soil-forming processes proceed in the same direction must also be mentioned. Then, instead of opposing each other, their effects are additive. The main difference between recent soils and older ones lies in the *intensity of the basic physicochemical process which is a function of time of development.* An old soil appears more "mature" than a young one. For instance, the intensity of eluviation and of acidification under temperate climate is always more marked in an old loam than in a young one (see Plates IX and X). The same may be said of the intensity of rubefaction and of the surface depletion of Fersialitic soils on terraces, which, under Mediterranean climate, is related to the age of the terraces (Plate XVI).

Complex profiles: soils with superposed parent materials

Soils developed from two superposed parent materials, an old one at depth and a young one (often more or less reworked) at the surface, are very frequently observed. These are *complex profiles* of which many examples will be found in the Atlas.

Two cases must be distinguished according to whether or not both parent materials have been influenced by subsequent genetic evolution.

The simplest case is of course that of a mere layering of materials with no genetic relationship. This can be observed when the surface material is very recent (or very thick). Often an abrupt boundary separates it from the old material which is frequently a Paleosol. The Paleosol is not always distinct, as it has usually been truncated by erosion before being buried, but has not undergone physicochemical modification (Profile II_4).

The much more complicated general case arises when older surface deposits are not thick enough to maintain the deeper Paleosol in a "fossil" state, out of the soil "profile." *A soil-forming process subsequent to the deposition of the surface layer may have taken place and simultaneously involved both layers of the heterogenous profile.*

In some circumstances, only "soil climate" is influenced by the deep layer, if it is impermeable. Because of their impermeability, "Glossic" Paleosols or those with a fragipan often preclude soil water drainage. The recent development of the surface horizons is more or less hydromorphic in nature depending on the prevailing bioclimatic factors. If evapotranspiration is high, hydromorphism is only slightly apparent. If on the contrary, it is low, hydromorphism is very marked, as a "perched water table" then develops above the fragipan (Profiles II_5 and XIII_2).

Finally, the deep layer very often acquires, through illuviation, certain fine particles or soluble elements from the surface layer. A complex B_t horizon then develops within the deep layer. The Beta horizon formed on limestone is a special case of this type of horizon (Profile II_6). Interpretation is most difficult under these circumstances as such soils are polycyclic and at the same time have complex profiles.

LEGENDS AND COMMENTS

PLATE I

Influence of general bioclimatic factors: climatic climaxes and analogous soils

(Profiles 1 and 2)

Elevation zones in mountains provide an excellent illustration of the concept of analogous soils. The subalpine level is characterized by evergreen forest with Ericaceae and by soil with a very thick acid mor. Depending on the nature of the parent material, soils are either Podzols or Lithocalcic soils with mor (see Profiles XI_6 and V_6). At lower elevation, at the humid montane level, deciduous trees begin to mix with evergreens and the podzolic development of soils is less pronounced.

PROFILE I_1. *Subalpine Podzol developed on terra fusca and reworked dolomitic moraine*: Cortina Forest, The Dolomites, Italy. Despite the large reserve of calcium and magnesium carbonates in the parent material, acidification is very intense. Below a 40-cm thick mor, an ashy-gray A_2 horizon (5-8 cm) and a reddish-brown spodic horizon (8-10 cm thick) are easily distinguishable.

REFERENCE: Prof. Susmel's Laboratory, Department of Forestry, Padua.

PROFILE I_2. *Podzolic Eluviated soil developed on reworked terra fusca* (karst pocket): Lente Forest, Vercors, Isère, France; elevation 1,000 m; fir mixed with beech. This example demonstrates the opposition that exists between the recent soil-forming process at the surface, which is podzolic, and the nature of the parent material, which is a Fersialitic Paleosol. Hence, this is a polycyclic soil. An acid moder at the surface lies above a bleached A_2 horizon (photo by Bottner).

NOTE. Bottner (1971) has shown how, in the high forests of Vercors, the evolution of the humus and of the upper part of the profile proceeds similarly from different parent materials. Under similar vegetation, these Podzolic Eluviated soils can be found overlying parent rock as widely varied as marl or crystalline rock. The latter profiles are then *monocyclic*. This group of soils is a good example of "analogous soils."

Site climaxes: influence of local ecological factors

(Profiles 3 and 4)

Two examples of extremely localized site climaxes are presented. One is affected by "soil climate," the other by the presence of *a tree* (sea-pine, *Pinus pinaster*) which has promoted very intense, but localized, podzolization.

PROFILE I_3. *Humic Lithocalcic soil with "hydromor" developed on calcareous talus*: at foot of horseshoe-shaped Creux du Van cliff, Swiss

Jura; cold exposure, elevation 1,140 m. Vegetation consists of *Lycopodio mugetum* with *Sphagnum* which is unusual, both at this elevation and on normally well-drained and aerated calcareous talus. Richard (1961) has demonstrated the role played by the exceptionally cold and wet soil climate, together with the protection afforded by the limestone wall, in the development of this type of soil.

PROFILE I₄. *"Egg-cup"- or "funnel"-shaped Podzol developed on Landes sand*: Cestas, Gironde, France. This Podzol has formed at the place of a very old, decayed sea-pine stump. Its taproot was inclined in the direction of flow of the ground water. This is reflected in the profile. The ashy-gray A₂ horizon occupies the place of the taproot and is bordered by a very hard, brownish-black aliotic layer. Within a few hundred years, a single tree has caused a locally well-developed Podzol to form from parent material very high in quartz (Landes sand). Yet neighboring profiles, having developed under deciduous forest, are Podzolic Eluviated soils.

REFERENCE: Juste, C., I.N.R.A. Laboratory, Pont de la Maye, Gironde, France.

Concept of parent material: Paleosols

(Profiles 5 and 6)

Most pre-Wurmian Quaternary formations have undergone ancient soil forming processes. These constitute Paleosols which on occasion become the "parent material" of current soils. Profiles I₅ and I₆ are good examples of Paleosols.

PROFILE I₅. *Cryoturbated Paleosol developed on ancient terrace*: Sainte-Hélène quarry, Meurthe-et-Moselle, France. Three superposed Paleosols are distinguishable. The silica-rich gravel of the terrace, *sensu stricto*, constitutes a "Tropical Ferruginous soil," probably dating back to Villafranchian. Above are two layers of loam filling *cryoturbated pockets*. One is old and strongly rubefied (Fersialitic soil), while the other is brown, more recent, less rubefied, and has generally been eliminated by erosion except in the central portion of pockets. In some sections of the forest, this second layer has received better protection and serves as support to "Glossic Eluviated soils" (Profile X₃; Le Tacon, 1966) (photo by M. Gury).

PROFILE I₆. *Detail of Paleosol*: Gave du Pau terrace, Livran, Pyrénées-Atlantiques, France. This photograph illustrates Icole's method (1973) of dating the parent material and of identifying the mode of weathering that is responsible for the formation of the Paleosol. The author used the degree of weathering of pebbles as his criterion. The kinds of secondary minerals present and the thickness of the weathered "cortex" of pebbles (which depends on the weatherability of the component rocks of pebbles) are good indices of the type and age of pedogenesis. Of course, judicious comparisons with terraces of known age are necessary.

PLATE 1

CLIMATIC AND SITE CLIMAXES, PALEOSOLS

LEGENDS AND COMMENTS

PLATE II

Polycyclic (polygenetic) soils

(Profiles 1-3; see also Profile I_2)

Profiles II_1-II_3, as well as I_2, are good examples of soils having undergone several "superimposed" evolution phases. Successive phases may result from different processes: climatic (weathering in Profiles I_2 and II_1), mechanical (cryoturbation in Profile II_1), or biological (humification in Profiles II_2 and II_3).

PROFILE II_1. *Eroded and cryoturbated terra fusca*: edge of the Bajocian limestone plateau of Haye, Meurthe-et-Moselle, France. Clay has formed through ancient decarbonation after weathering of the hard limestone. This material is "fersialitic" in nature, very rich in iron and sometimes in alumina (see also Profile IX_4). It may have undergone very localized and recent genetic processes, according to whether it constitutes a "pocket" of cryoturbation or whether, on the contrary, it forms a thin reworked layer mixed with frost-cleft fragments of limestone. There is formation of a carbonate-free, polycyclic "Calcic Brown soil" in the first instance and, in the second, of a "Brunified Rendzina" having undergone secondary recarbonation.

Profiles II_2 and II_3 have both been subject to successive humification processes, though under quite different ecological conditions.

PROFILE II_2. *Brunified Isohumic soil*: Limagne peatland, Puy-de-Dôme, France. The humus in this profile is highly polymerized and, in the past, underwent a climatic maturation similar to that of Chernozems. It would date back to the Atlantic period and could have formed during a relatively dry period due to the barrier afforded by the Dôme Mountains. Because of its exceptional resistance to biodegradation, this very dark and stable humus has remained intact for several millennia. Under the influence of current vegetation, a forest "mull" with much faster turnover has developed at the surface and is the expression of a "brunification" process characteristic of the current Atlantic climate.

REFERENCES: *Soil Map of France*, Vichy sheet, Prof. Servat, Soils Laboratory, Montpellier, France; Gachon, 1963.

PROFILE II_3. *Podzolic Ochric soil with black B_h horizon*: Aigoual Massif, Gard, France. This profile is situated below the subalpine meadow, at the level of the montane beech-grove, on crystalline schist parent rock. The top portion of the profile is a "Podzolic Ochric soil" with an *ocher* B_s horizon, high in lightly colored fulvic acids. This soil is characteristic of the beech-grove. However, at a greater depth, a *black spodic* B_h horizon (particularly high in free aluminum) is observed. Its presence cannot be explained at first. According to Warembourg et al. (1973), this could be an old "Cryptopodzolic Ranker" formed in the meadow and with deep humification. It features a very dark A_1B_h horizon. The black B_h horizon is a relic of the meadow soil. The return of the beech-grove appears to have modified the features of the top of the spodic horizon. It is also likely that the upper part of the profile has been somewhat reworked mechanically.

Complex and polycyclic soils

(Profiles 4-6)

These are soils that have developed in superposed materials, each having a different "age" and thus a different "history."

PROFILE II₄. *Brown soil developed above a Red Paleosol*: Broue quarry, Nogent-le-Roi, Eure-et-Loir, France. This profile shows the superposition of two distinct layers of different ages but without any true "genetic link." A recent sandy loam (Brown soil) lies on top of a Red Paleosol (rotlehm with discolored glossic bands) that has been truncated by erosion. Truncation is illustrated by the stone line at the base of the Brown soil.

REFERENCE: *Seminar on cartography, 1965*, Association française pour l'Etude du Sol, E.N.S.A., Grignon, France.

PROFILE II₅. *Acid Brown soil developed above Glossic Paleosol*: Charmes Forest, Vosges, France (Becker, 1971). This complex profile has similar origin and composition. The Acid Brown soil formed recently at the surface lies on top of a non-rubefied Glossic Paleosol. However, two peculiarities must be stressed. Examination of microstructures reveals that the brunified surface layer has been reworked by cryoturbation. Furthermore, the slightly porous and slowly permeable glossic horizon often determines the degree of *hydromorphism* of the upper part of the profile. In the present case, very high evapotranspiration rates, due to a dense forest cover, oppose this effect which remains small. But any degradation of the forest intensifies hydromorphic conditions by increasing the duration and thickness of perched water tables. Examples of this will be found in Profiles X₃ and XIII₂ (photo by Becker).

PROFILE II₆. *Complex Eluviated Brown soil with β horizon*: Pont-à-Mousson region, Meurthe-et-Moselle, France. The history of this profile is particularly complicated. It has developed from two materials. A layer of soliflucted loam lies on top of a clay layer formed from ancient decarbonation of Bajocian limestone. Evidence of solifluction is found in the few hard limestone pebbles that have been carried to the surface of the loam. The deep reddish-brown clay layer has been subsequently enriched with fine clay by illuviation from the upper loam layer. This constitutes the β horizon. This process has been reported by several authors (Bartelli and Odell, 1960; Ducloux, 1970; Robin and De Coninck, 1975).

PLATE II

Polycyclic and Complex Soils

CHAPTER II

IMMATURE SOILS AND SOILS WITH LITTLE HORIZON DIFFERENTIATION

In several taxonomies, soils with an AC profile are grouped in a single class. Such soils, even if very humic (dark and thick A_1 horizon), are all characterized by the absence or the weak development of a brown (B) or B_v horizon of alteration. This practice results in classifying together soils whose ecology and degree of evolution are very different.

In some cases, absence of a (B) horizon is indeed indicative of a young profile. Under particularly favorable conditions, organic horizons develop very rapidly. A large incorporation of organic matter is not incompatible with low weathering of the mineral material. This is particularly the case of *deposited Colluvial and Alluvial soils*. These soils can actually be considered as "immature."

Under different conditions, however, the weak morphological differentiation of the profile, which appears as an A_1C profile, may actually mask a high degree of transformation of both organic and mineral matter. In particular, special kinds of organomineral complexes are formed which characterize two fundamental processes—"cryptopodzolization" (influenced by strongly acid organic matter at high elevations) and "andosolization" (on volcanic material).

It is my opinion that these two important categories of soils should be clearly separated into two different classes:

(i) *Deposited* (or eroded) *immature soils*: Under certain conditions, these soils may accumulate organic compounds or even begin to evolve towards other classes.

(ii) *Desaturated humic soils with little horizon differentiation*: These soils are formed under humid climate (usually high altitude climate) which promotes a high rate of weathering and a strong acidification of the organic matter. This is accompanied by the formation of iron and aluminum organomineral complexes which may even undergo local redistribution within the profile. All these processes are marked by the presence of a thick, uniformly black A_1 horizon (Ranker-Andosols).

For reasons of genetic affiliation, young phases (Rankers of erosion-Vitrisols) have been grouped with their mature homologues (Cryptopodzolic Ranker-Andosols).

ERODED OR DEPOSITED IMMATURE SOILS (A₁C profile)

Soils whose weak development is due to "mechanical" causes. Under any type of climate, the profile is rejuvenated either by erosion, whereby a weakly weathered parent rock is periodically exposed, or by deposition of a young material.

A *deposited soil* can either be classified according to the nature of the material and its mode of transport (eolian, alluvial, or colluvial) or according to its *incipient pedogenic development*, if the deposit is old enough for such a development to have begun. In this case, the soil would form a transition towards another class. Examples of this are Brunified soils, Hydromorphic soils, Calcimagnesian soils (if the material is very calcareous), etc.

Two groups will be distinguished according to mode of transport and geomorphological conditions:

(i) *Alluvial soils*, typically displaying wide fluctuations of the water table and weak reducing conditions. Such soils are found in the floodplains of rivers.

(ii) *Colluvial soils* of usually coarser materials which have been removed from slopes by erosion. They generally have no water table.

In reality, many Alluvial and Colluvial soils exhibit a more or less marked evolution towards other classes, either because of interruption of deposition or, in the case of Alluvial soils, because of lowering of the water table or changes in its redox potential. Wide fluctuations of the water table can, furthermore, promote an important growth of vegetation and a deep penetration of humus. The humus then undergoes a vertic or chernozemic type of "maturation." A whole range of intergrade soils can be found under these conditions, such as Brunified, Isohumic, Rendzinic, Vertic, or Gleyed Alluvial or Colluvial soils (see Table 1 for details). Some of these soils are grouped with Alluvial soils; others, on the contrary, are classified with the soils to which they are related.

NOTE 1. As a rule, since they were transported, these soils are characterized by *loose* and *weakly weathered parent material*, thus with no or little iron coloration. They are usually of grayish color. However, if before being transported, the materials had undergone an earlier pedogenic development, they may be relatively rich in free iron and appear brown. This is not the result of a modern pedogenic process but of a simple "inheritance." Such a soil would be considered "immature" so long as a *stable* "structure," characteristic of a (B) horizon, has not developed.

NOTE 2. Soils that are weakly developed due to climatic conditions, such as very cold climate Toundra soils and Cryosols or very dry climate Desert soils, are not presented in the Atlas.

DESATURATED HUMIC SOILS WITH LITTLE HORIZON DIFFERENTIATION (TABLE 2)

Generally acid, desaturated soils with dark, often black, color and no apparent horizon differentiation, simulating an AC profile. The humic A₁ horizon may be very thick.

These soils form under humid, often montane, climate, which promotes organic matter accumulation at various stages of humification. They develop from noncalcareous rock. Homologous soils developed from calcareous parent material are usually classified with the Calcimagnesian soils (Chapter III) even if the most organic among the latter reveal a certain evolutionary convergence (case of high elevation humic soils).

Rankers develop from crystalline aluminous-siliceous rock of various kinds. *Andosols* are formed in vitreous volcanic parent material.

Rankers

The *Ranker of erosion* (or Slope Ranker) occurs when the steepness of the slope prevents the formation of a (B) horizon of alteration. The A_1 horizon consists of weakly transformed and weakly humified organic matter and forms an abrupt boundary directly with the hard rock. The high-altitude *Alpine Ranker* is similar, except that the A_0 horizon is thicker and more humified.

The high-altitude *Cryptopodzolic Ranker* (or "Atlantic" Ranker along the Atlantic coast) is a *mature* soil but with *weak horizon differentiation*. The dark and uniform A_1 (or A_h) horizon results from the evolution of amorphous organomineral complexes, like the B_h horizons of Podzols (the profile is therefore designated A_1B). This type of soil is often classified with Podzolized soils under the heading "Humic Cryptopodzolic soil." It is frequently found on horizontal landscape or in depressions with poor drainage, with meadow or moorland short vegetation. Under forest cover at higher elevations, it evolves towards a Humic Ochric soil.

Andosols

Like Cryptopodzolic Rankers, Andosols are very rich in highly humified organic matter which is redistributed at depth. Two factors are responsible for their formation: (i) a *climatic* factor—humid climate with no dry season, often montane; (ii) a *site* factor—volcanic mineral parent material high in vitreous elements. The massive liberation of "allophane" (imperfectly crystallized aluminosilicates) and of amorphous products, which is characteristic of these soils, results from hydration and rapid weathering of these noncrystalline materials. The very abundant amorphous mineral compounds (alumina, silica, iron hydrates) immobilize the soluble prehumus substances rapidly. They do not migrate except in transition soils.

Vitrisols are *young* Andosols developed from weakly weathered and weakly hydrated volcanic parent materials. They are thus high in sand or vitreous loam, yet still low in allophane.

Typical **Andosols**, which have high allophane and amorphous mineral contents, display no other evolutionary feature than a considerable accumulation of organic matter. They are found mainly at high altitude where the very humid or cold climate, or both, precludes any intense and prolonged drying of the profile. However, under constant humid climate (in some equatorial regions), they can also be found in plains. In a given mountain, typical Andosols usually characterize a well-delineated *climatic elevation zone*. Two kinds of Andosols can be noted:

(i) *Undifferentiated Humic Andosols* with no clear (B) horizon, or else whose (B) horizon is strongly impregnated with humus. They derive from strongly basic and rapidly weathered volcanic ash or pumice.

(ii) *Differentiated Andosols* (sometimes called "Chromic Andosols") with a brown or ocher (B) horizon which is less humic and clearly distinct from the A horizon. They are found more frequently on consolidated volcanic rock or on any other parent material with slower weathering than in (i).

Evolution of Andosols. Many Andosols undergo an evolution superimposed on andosolization which brings them closer to other classes of soils. Their andic character *per se* is less emphasized. Examples are *Brunified Andic soils* (temperate climate) and *Ferralitic Andic soils* (warm climate). Both of these occur at lower elevation and are affected by a more or less intense seasonal period of desiccation.

Elevation sequences of Andosols

Above the Andosol elevation zone, a whole range of intergrade soils is influenced by local ecological conditions. There are: *Andic Rankers* at the short vegetation elevation zone above the forest zone; *Andopodzolic soils,* in which acidification is tied to climate and nature of parent material; and *Hydromorphic Andic soils,* which are waterlogged and in which iron is reduced.

Below the Andosol elevation zone, the dry season becomes more pronounced and only "Andic"* soils are found. These are lower in allophane but higher in halloysite or kaolinite clay minerals which derive from allophane by neoformation. There are *Brunified Andic soils* under temperate climate, and *Ferralitic Andic soils* under equatorial climate. The latter differ from the former by a more intense desilicification and by formation of gibbsite.

---

*The reader is reminded that, in the French system of soil classification, an adjective denotes a lower stage of development than a noun.

Table 1.   Evolution of Deposited Immature Soils

I. ALLUVIAL SOILS

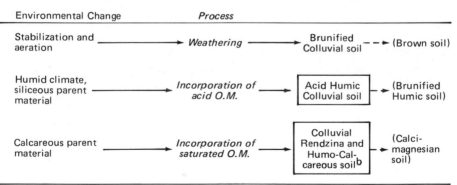

Environmental Change

Evolution of the weakly weathered and slightly humic ⟨ Alluvial soil ⟩ :

Environmental Change	Process		
Increase in aeration (cessation of deposition)	→ *Weathering* →	Brunified Alluvial soil	→ (Brown soil)
More severe reducing conditions with little fluctuation of the water table	→ *Weathering (+ partial reduction)* →	Alluvial Gley soil[a]	→ (Slightly Humic Gley soil)
Widely fluctuating water table with alternate wet and dry periods { Calcareous material	*Humification and structure development* →	Rendzinic Alluvial soil	→ (Calcimagnesian soil)
Clay-rich material	*Formation of stable clay-humus complexes* →	Vertic Alluvial soil	→ (Vertisol)

II. COLLUVIAL SOILS

Environmental Change	*Process*		
Stabilization and aeration	→ *Weathering* →	Brunified Colluvial soil	- - → (Brown soil)
Humid climate, siliceous parent material	→ *Incorporation of acid O.M.* →	Acid Humic Colluvial soil	→ (Brunified Humic soil)
Calcareous parent material	→ *Incorporation of saturated O.M.* →	Colluvial Rendzina and Humo-Calcareous soil[b]	→ (Calcimagnesian soil)

[a]Studied under Hydromorphic soils (Chapter VII).
[b]Studied under Calcimagnesian soils (Chapter III).

Table 2.   Evolution of Soils with Little Horizon Differentiation

I. HARD AND ACID CRYSTALLINE ROCK

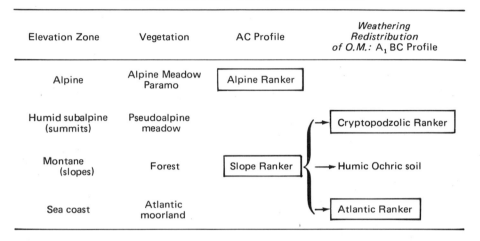

Elevation Zone	Vegetation	AC Profile	Weathering Redistribution of O.M.: A_1 BC Profile
Alpine	Alpine Meadow Paramo	Alpine Ranker	
Humid subalpine (summits)	Pseudoalpine meadow		Cryptopodzolic Ranker
Montane (slopes)	Forest	Slope Ranker	Humic Ochric soil
Sea coast	Atlantic moorland		Atlantic Ranker

II. VOLCANIC ROCK

1. Humid Climate with No Dry Season

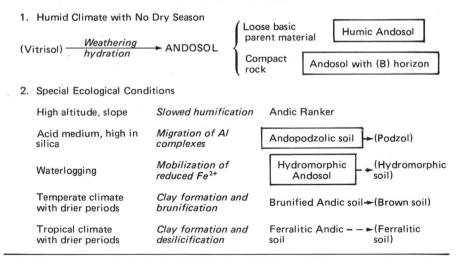

(Vitrisol) $\xrightarrow{\text{Weathering hydration}}$ ANDOSOL

Loose basic parent material → Humic Andosol

Compact rock → Andosol with (B) horizon

2. Special Ecological Conditions

High altitude, slope	*Slowed humification*	Andic Ranker
Acid medium, high in silica	*Migration of Al complexes*	Andopodzolic soil → (Podzol)
Waterlogging	*Mobilization of reduced Fe^{2+}*	Hydromorphic Andosol → (Hydromorphic soil)
Temperate climate with drier periods	*Clay formation and brunification*	Brunified Andic soil → (Brown soil)
Tropical climate with drier periods	*Clay formation and desilicification*	Ferralitic Andic soil – – ► (Ferralitic soil)

III₁: ALLUVIAL SOIL

(Sol alluvial)

F.A.O.: *Eutric Fluvisol*; U.S.: *Psammentic Udifluvent*

Location: Along Alsace Canal, Nambsheim, Haut-Rhin, France.
Topography: Level. Elevation 200 m. Undulating microrelief.
Parent material: Recent alluvium of the Rhine River.
Climate: Temperate continental. P. 510 mm; M.T. 10.9°C (at Colmar).
Vegetation: Fallow, calcicole shrubs.

Profile Description

A₁ (0-10 cm):   Grayish-brown (10 YR 5/2) slightly humic mull, sandy loam; common grass roots; clear boundary.
C₁ (10-55 cm):   Light brownish-gray (10 YR 6/2) sandy loam; single-grained to fine crumb structure, loose; effervescent; few roots; gradual boundary.
C₂ (55-75 cm):   Light gray loamy sand; single-grained structure; effervescent; many roots above water table.

NOTE. Before the Alsace Canal was dug a few years ago, causing the water table to fall, the water table fluctuated from the surface to a depth of about 1 m.

Geochemical and Biochemical Properties

Particle-size distribution: Relatively homogeneous sandy loam down to the 55-cm depth; 7-8% silt, 43-45% sand, and 28% $CaCO_3$. Sandier (58% sand) from 55 to 75 cm. These variations express differences in alluvial deposits.

Organic matter and exchange complex: The mull is not yet very humified (4.5% O.M. in A₁) because of the weak development of the profile. C/N ratio of 11.8 indicates high biological activity. Organic matter content decreases gradually with depth (1% near 50-cm depth). The exchange complex is 100% saturated because of the presence of carbonates and pH values of 8 in A₁ and 8.4 in the mineral horizons.

Free iron: Horizons C₁ and C₂ are low in free iron (0.25%). A slightly higher content (0.35%) in the surface horizon indicates very moderate weathering induced by the humus.

Genesis: Very immature soil. The pale color and the small quantities of clay and free iron indicate low weathering of minerals. No migration is evident, not even of carbonates whose content is approximately constant throughout the profile. Variations in particle size are simply due to differences in the deposits, which is a constant feature of Alluvial soils. The periodic flooding of this soil maintains it at a low stage of development (Kubiena's *paternia*, 1953).

At slightly elevated positions from the riverbed (along the bank shoulders) and shielded from flooding, the Alluvial soil is decarbonated and brunified. In lateral depressions beyond the bank shoulders, alluvium is finer in texture and subjected to a water table with small fluctuations and slow circulation. There, Alluvial Gley soils are found (Profile XIV₁).

REFERENCE: Unpublished data from the Agricultural Research Station, Colmar, 1973.

III₂: BRUNIFIED ALLUVIAL SOIL

(Sol alluvial brunifié)

F.A.O.: *Eutric Fluvisol*; U.S.: *Psamment*

Location: Seurre, Côte-d'Or, France.
Topography: Level alluvial valley. Elevation 180 m.
Parent material: Recent alluvium of the Saône River.
Climate: Temperate Atlantic. P. 720 mm; M.T. 10.5°C.
Vegetation: Natural grassland.

Profile Description

A₁ (0-5 cm):   Brown sandy mull.
A(B) (5-80 cm): Light yellowish-brown (10 YR 6/4) loamy sand; single-grained structure, loose.
B$_g$ (80-100 cm): Reddish-yellow (7.5 YR 6/6) sandy loam; loose, small brownish-black concretions.
C: Very light-colored sand with mottles.

Geochemical and Biochemical Properties

Particle-size distribution: Homogeneous sandy texture; 80% sand at the surface, 75% at 80-cm depth where clay content increases slightly (6.5%).
Exchange complex: Traces of carbonates; pH 7. Saturated exchange complex.
Biochemistry: Low organic matter content at the surface (4.5%), decreasing rapidly with depth. C/N ratio is 10-11, indicating intense biological activity.

Genesis. Relatively immature soil but slightly brunified. At depth, hydromorphic features are appearing with iron moving up and precipitating in a rusty, ocherous band (see photograph) at the upper limit of the capillary fringe. The brown color of Alluvial soils can either originate in incipient weathering (this is true "brunification") or occur simply because the transported material was originally brown. This is mainly the case here. It is also a general characteristic of alluvium of the Saône and of many other French rivers.

As in the case of all Alluvial soils, texture varies abruptly, either vertically from one horizon to the next (not the case here), or horizontally from one profile to the next. A few kilometers from this site, in the same valley, is a similar profile but much higher in clay (40% clay). Such a high clay content (together with bound iron) is, of course, inherited from the parent material. Thus, a prismatic structure develops rapidly due to "shrinkage" of the clay during dry periods. None of this demonstrates prolonged development. It would appear that, although it simulates a "well-developed" brown profile, this soil should still be classified with immature soils for genetic reasons.

REFERENCE: *Soil Map of France (1:100,000)*, Dijon sheet, Agricultural Research Station, Dijon, 1973.

III$_3$: VERTIC ALLUVIAL SOIL

(Vertic Chernozem with gley)

(Sol alluvial vertique; chernozem vertique à gley)
F.A.O.: *Calcic Chernozem*; U.S.: *Vertic Calciaquoll*

Location: South of Olginskia, 20 km southeast of Rostov, U.S.S.R.
Topography: Level alluvial valley (flooding). Elevation 2-3 m.
Parent material: Recent clayey alluvium, resting on sandier alluvium.
Climate: Dry continental. P. 400-450 mm; M.T. 9°C (Jan. −5°C, July +24°C).
Vegetation: Wet grassland, vegetable crops, or rice paddies.

Profile Description

A$_1$ (0-20 cm):   Very dark gray (10 YR 3/1) mull; crumb structure; gradual boundary.
A$_1$(B) (20-50 cm):   Black (10 YR 2/1) clay, almost constantly wet; prismatic to blocky structure, vertical cracks; gradual boundary.
(B)C (50-160 cm):   Dark gray (10 YR 4/1) clay, becoming dark grayish-brown (10 YR 4/2) at depth; prismatic to massive structure, irregular powdery calcareous spots; clear boundary.
IICG$_0$ (160 cm+):   Loamy sand; water table fluctuation zone; mottles alternating with greenish gray reduced zones.

Geochemical and Biochemical Properties

Particle-size distribution and clay minerals: Two materials are present. Clay alluvium, 160 cm thick [over 50% clay in A$_1$(B)], rests on a loamy sand with only 12% clay. Most of the clay minerals are inherited from the alluvial material and consist of interstratified illite-montmorillonite minerals. The rest consisting of montmorillonite is probably neoformed.

Exchange complex: Very high exchange capacity (40 meq/100 g) due to the nature of the clay minerals and humus. The exchange complex is saturated mainly with Ca^{2+} and Mg^{2+} in a 2/1 ratio and traces of Na$^+$. The upper 45 cm of the profile are decarbonated, but 2-4% CaCO$_3$ is present below this depth. Although the ground water contains 0.7% salt as NaCl and CaSO$_4$, the soil is only slightly saline due to frequent flooding with fresh water.

Biochemistry: Very definite "isohumic" feature with 6% organic matter in A$_1$ and 3-4% in A$_1$(B). Organic matter content is higher than in a typical Vertisol.

Genesis. This soil is often called "meadow soil" (*Wiesenboden*). While this name provides a good indication of its ecology, it does not satisfy the pedologist. This soil is indeed "alluvial," considering the origin of the parent material (alluvial deposit) and the occurrence of periodic flooding. However, wide fluctuations of the water table due to a zonal temperate steppe climate promote a deep incorporation of highly matured organic matter, vertic features at the surface (shrinkage cracks), and iron segregation at depth (gley). This is not a true Vertisol, because the soil is saturated for too long, but a Vertic Alluvial soil. The name "Vertic Chernozem with gley" is also acceptable.

REFERENCE: *Guidebook to the Volga-Don Excursion*, 10th International Congress of Soil Science, Moscow, 1974.

III₄: HUMIC COLLUVIAL SOIL

(Sol colluvial humifère)

F.A.O.: *Humic Cambisol*; U.S.: *Typic Haplumbrept*

Location: Morthomme Woods, Fossard Forest, Vosges, France.
Topography: Foot of steep slope, southeast exposure. Elevation 570 m.
Parent material: Remiremont granite and gneiss colluvium.
Climate: Lower montane. P. approx. 1,600 mm; M.T. approx. 8°C.
Vegetation: Mixed forest with fir, spruce, oak, sycamore maple. Mull flora consisting of *Asperula odorata, Mercurialis perennis*.

Profile Description

A_1 (0-25 cm):   Very dark grayish-brown (10 YR 3/2) very humic mull; medium crumb structure, irregular and friable; very porous and aerated; contains much angular gravel (85%); many roots of all dimensions; gradual boundary.

A_1(B) (25-60 cm):   Dark brown to brown (7.5 YR 4/2) mull; crumb structure, friable; contains much angular gravel (90-95%); common roots of all dimensions.

Geochemical and Biochemical Properties

Particle-size distribution: Fine earth is a sandy loam with about 15% clay. Particle-size distribution is homogeneous throughout the profile.

Exchange complex: Relatively high exchange capacity [24 meq/100 g in A_1, 16 meq/100 g in A_1(B)] due to abundance of organic matter. The S value is 3-4 times higher than in neighboring Acid Brown soils formed on the same parent material [6 meq/100 g in A_1, about as much in A_1(B)]. S/T is 25% in A_1 (pH 5.6) and 36% in A_1(B) (pH 6.4). This mull is almost "mesotrophic."

Biochemistry: Abundant, well-humified organic matter [9.6% in A_1, 3.5% in A_1(B)]. C/N ratios of 17 in A_1 and 10.5 in A_1(B) are characteristic of a fairly active mull under evergreen cover.

Iron and aluminum hydroxides: Like clay, free iron (1.3%) and free aluminum (0.8%) are uniformly disseminated. Such high contents are indicative of relatively strong weathering of the gravel, but especially of an important lateral supply.

Genesis. This Colluvial soil (situated below a Slope Ranker) displays certain "juvenile" characteristics due to the rapid renewal of material. Clay and iron are distributed uniformly. The soil acquires solid elements (clay) and soluble compounds (organomineral complexes supplying cations) from the Slope Ranker. This promotes humification and incorporation of matured organic matter to a great depth. High iron and aluminum contents indicate that "brunification" is beginning. This process is obviously accelerated by lateral supplies. Although the soil is acid, exceptional conditions (aeration; availability of water, nitrogen, and nutrients) contribute to the growth of a demanding flora, yet tolerant to acidity and aluminum (*Mercurialis perennis*, for instance).

REFERENCE: Gury, M., and C. Rodriguez, *Cartographic study of the Fossard Forest.* C.N.R.S. (*unpublished*).

III₅: SLOPE RANKER

(Ranker de pente)

F.A.O.: *Ranker*; U.S.: *Lithic Haplumbrept*

Location: Fossard Forest, Vosges, France.
Topography: Very steep slope, east exposure. Elevation 610 m.
Parent material: Remiremont granite.
Climate: Humid montane. P. approx. 1,600 mm; M.T. approx. 8°C.
Vegetation: Sparse forest of spruce and fir. *Calluna vulgaris*, *Vaccinium myrtillus*, and moss on forest floor.

Profile Description

A₀F (0-5 cm): Dark reddish-brown (5 YR 3/2) humic horizon, fibrous; many roots; gradual boundary.

A₀H (5-20 cm): Black (5 YR 2/1) fine hydromorphic humus; fine granular structure when dry; sticky when wet; few coarse quartz grains; common roots; abrupt boundary with bedrock.

R: Unweathered granite forming a very hard slab.

Geochemical and Biochemical Properties

Organic matter and particle-size distribution: Mor containing small amounts of mineral material; 62% organic matter in A₀H. Mineral impurities consist of 10% sand and 15% clay, and probably come from the weathering of micaceous minerals that were incorporated within the humus. C/N ratio of 30.4, characteristic of a very acid mor. Extractable carbon is 35% (pyrophosphate followed by 0.1N NaOH extraction).

Exchange complex: Very high exchange capacity (75 meq/100 g). This is, nevertheless, a low value considering the amount of organic matter present. A low value of S (less than 2 meq/100 g) gives an exceptionally low S/T percentage (2.4%) and a pH of 3.5.

Iron and aluminum hydroxides: Both iron (0.35%) and aluminum (0.46%) contents are low, because of their biogeochemical origin and because of depletion, as soluble complexes are leached out laterally. The aluminum/clay ratio is around 0.03. This suggests a moderate degree of biochemical weathering which is probably underestimated.

Genesis. This Slope Ranker contrasts with a "Cryptopodzolic Ranker" which features a more or less well-expressed, spodic-type horizon at the base of the A₀A₁ horizon. This is linked to the steep slope. Erosion eliminates most mineral products of weathering. The pseudosoluble complexes formed in A₀ are leached out laterally. The profile becomes depleted of Fe, Al, Ca, and Mg cations. Thus, no mixed or mineral (B) or B horizon can form. Acidification of the profile proceeds rapidly. Humification and mineralization of the organic matter are slowed by the extreme acidity and by unfavorable soil climate conditions (cold and near permanent wetness due to runoff of precipitation down the slope). A *very acid hydromor* develops.

REFERENCE: Gury, M., and C. Rodriguez, *Cartographic study of the Fossard Forest. C.N.R.S. (unpublished).*

III_6: ALPINE RANKER

(Ranker alpin)

F.A.O: *Humic Cambisol*; U.S.: *Lithic Humitropept*

Location: Laguna Negra, Paramo de Sumapas, Cundimarca, Colombia.
Topography: 12-25% slope. Elevation 3,700 m.
Parent material: Consolidated sandstone.
Climate: Tropical *alpine*. P. 1,800 mm; M.T. 4°C.
Vegetation: "Paramo" grasses, *Sphagnum*, *Speletia* sp.

Profile Description

A_0 (0-30 cm):   Very dark grayish-brown (2.5 Y 3/2) holorganic horizon, hydromor,
black when moist; granular to blocky structure; nonsticky, nonplastic;
very porous; constantly high water content; medium and fine roots
increase in number with depth, matted at lithic contact; abrupt boundary.

R: Light gray to white consolidated fine quartzous sandstone; no apparent
surface weathering.

Biochemical Properties

General composition: Approximately 74% organic matter in A_0, mixed with small
amounts of quartz sand.

Exchange complex: High exchange capacity, greatly exceeding 100 meq/100 g.
However, saturation percentage is very low (3.3%) with Mg^{2+} and K^+ as the dominant
cations. pH (water) 4.2. Al^{3+} is exceptionally abundant (18.4 meq/100 g).

Humification: Grasses and moss supply the organic matter; therefore, the C/N ratio is
low (15). Although the organic matter is not bound to mineral matter for lack of adsorbing
clay minerals, it is nevertheless fairly well decomposed. Successive extractions (with
pyrophosphate, followed by 0.1N NaOH) yield 49% extractable carbon. Moreover, a FA/HA
ratio of 0.3 indicates that bioclimatic transformation of humus compounds is taking place.

Genesis. The genesis of this slightly peaty (hydromor) Alpine Ranker is
conditioned by its topographic position (slope) and by a cold and humid climate of
altitude, which promotes the formation of a thick organic horizon. A (B) horizon
of alteration cannot develop at this elevation because weathering proceeds too
slowly to adequately balance rejuvenation by erosion. Like the Slope Ranker, this
is a *true Ranker*, quite different from a "Cryptopodzolic" Ranker. However, the
evolution and properties of the humus, or "alpine mor," contrast with those of the
Slope Ranker (Profile III_5). A much lower C/N ratio points to the influence of the
original plant material (grasses, instead of Ericaceae). Furthermore, seasonal
climatic variations are more marked. Thus, "humification" progresses further than
in the Atlantic Slope Ranker. Extractable carbon is close to 50%. This is relatively high in the absence of clay minerals which normally promote humification.

REFERENCE: Faivre, P., Codazzi Institute, Bogota, Colombia. Organic matter
study by F. Andreux.

PLATE III
ALLUVIAL AND COLLUVIAL SOILS, RANKERS

IV₁: SUBALPINE CRYPTOPODZOLIC RANKER
(Humic Cryptopodzolic soil)
(Ranker cryptopodzolique subalpin; sol cryptopodzolique humifère)

F.A.O.: *Leptic Podzol*; U.S.: *Typic Cryorthod*

Location: Haut de Falimont, Vosges, France.
Topography: Summit of granite dome. Elevation 1,300 m.
Parent material: Granite and more or less cryoturbated arenaceous weathering products of granite.
Climate: Humid subalpine. P. 2,000 mm; M.T. 4°C.
Vegetation: Primary meadow of "Hautes Chaumes"* with *Nardus stricta*, *Festuca rubra*, *Deschampsia flexuosa*, *Anemone alpina*, *Vaccinium myrtillus*, *V. vitis-idaea*.

Profile Description

A_0A_1 (0-7 cm): Black organic layer, fibrous; many roots.
A_1 (7-15 cm): Very dark gray (10 YR 3/1) organic horizon, with bare quartz grains; massive structure; sticky when wet, soft when dry; common fine roots; gradual boundary.
A_1B_1 (15-40 cm): Very dark brown (10 YR 2/2) loamy sand, becoming lighter with depth; fine granular structure, friable; common fine roots; gradual boundary.
A_1B_2 (40-60 cm): Yellowish-brown with darker organic matter pockets.
BC: Light yellowish-brown gravelly sand with weathered pebbles.

Geochemical and Biochemical Properties

Particle-size distribution: Loamy sand with 10.7% clay at 15 cm, decreasing to 5.7% in BC. The profile is thus young. Weathering is most intense at the top and compensates for eluviation of clay.

Exchange complex and pH: Exchange capacity is fairly high (22 meq/100 g in A_1B_1) due to the high content of well-humified organic matter. *S/T is low throughout the profile*, which is a property of this type of soil (4% in A_1B_1 and 8.7% in BC). pH (water) increases from 4 in A_1 to 4.8 at depth.

Biochemical properties: Very high organic matter content, decreasing with depth (from 15% in A_1, to 8% in A_1B_1, to 1.1% in BC). C/N ratio (13-15) is low because of the dominance of grasses and indicates intense biological activity during certain seasons. A high proportion of extractable carbon (52%) in A_1B_1 is the sign of a "spodic" feature.

Iron and aluminum: The high iron content, which decreases with depth, plays an active role in immobilization of organic compounds (1.9% in A_1, 1.4% at 40 cm). Aluminum is also high and migrates slightly (0.7% in A_1, 1% at 40 cm). The aluminum/clay ratio is above 0.1 in A_1B_1, which shows that clay minerals are being degraded (cryptopodzolization).

Genesis. This is a humic soil found on summits and exposed ridges of the Atlantic mountains, where a "primary" meadow with subalpine features prevails. Due to the cold and humid climate of altitude, this profile displays cryptopodzolic characteristics, such as humic compounds in the AB horizon resulting from *insolubilization* of prehumus substances, base desaturation, and high release of aluminum. However, it is morphologically not well differentiated. Hence, it is akin to the Slope Ranker with which it may form any number of transitions (Franz, 1956). However, contrary to the Slope Ranker, this soil is subjected to "bioclimatic" evolution. The intense biological activity during the spring is not incompatible with subdued podzolization. This soil becomes "brunified" [ochric (B) horizon] when under subalpine beech-grove vegetation, but its initial humic character remains (Profile II₃; Warembourg et al., 1973).

REFERENCE: *Soil Map of France (1:100,000)*, Saint-Dié sheet, C.N.R.S., 1973.

*Translators' note: *Hautes Chaumes* refers to areas of the Vosges mountains characterized by pasture grounds on often calcareous plateaus.

IV₂: ATLANTIC RANKER

(Ranker atlantique)

F.A.O.: *Leptic Podzol*; U.S.: *Humic Haplorthod*

Location: Barquero, Galicia, Spain.
Topography: Coastal platform with gentle northwest slope. Elevation 100 m.
Parent material: Granite.
Climate: Atlantic with Mediterranean trend. P. 1,400 mm; M.T. 10°C.
Vegetation: Atlantic moorland with Ericaceae and grasses.

Profile Description

A_0A_1 (0-15 cm): Black very humic horizon, fibrous; many roots.
A_1 (15-40 cm): Dark reddish-brown (5 YR 2/2) quartz grains coated with humus; irregular coarse crumb structure, loose when dry; diffuse boundary.
A_1B (40-75 cm): Dark grayish-brown (10 YR 4/2) becoming lighter with depth; "fluffy" structure, friable.
(B)C (75-90 cm): Grayish-brown (10 YR 5/2) gravelly sand; single grains tinted with rust.
R: Hard granite.

Geochemical and Biochemical Properties

Rankers developed on granite of the coastal zone of Galicia are very homogeneous. Their textural, chemical, and mineralogical properties are similar to those of Subalpine Rankers. Texture is loamy sand. Clay content (around 10%) is fairly constant throughout the profile. Because the granite parent material is more acid than in the profile of the Vosges (Profile IV₁), iron content is lower, whereas aluminum content is unchanged (1%) and varies little throughout the profile. Base saturation is low, like in high-elevation Rankers (S/T varies from 6 to 8%; pH 4.5-5).

Biochemical properties: It is instructive to compare these properties to those of the Subalpine Ranker. Here too, the C/N ratio (13.5-15) demonstrates high seasonal biological activity. Very high organic matter content varies between 17% (A_0A_1) and 12% (A_1B). Because extractable carbon is 68%, the cryptopodzolic character of the A_1B horizon is even more marked than in the Ranker of the Vosges. But, contrary to high elevation Rankers, humic acids are more abundant than fulvic acids. This is due to the polymerizing effect of a somewhat drier warm season.

Genesis. Because of their "cryptopodzolic" development, Atlantic Rankers, as described by Franz (1956), are very similar to Subalpine Rankers. They are thicker and humification is more advanced. Their development is influenced by comparable ecological conditions. *Vegetation* consists of short moorland with shallow roots and *parent material* is made of granite containing weatherable minerals. (NOTE: A Podzol has developed on neighboring quartzite.) Only climate is different. But due to the proximity of the sea, high humidity and low potential evapotranspiration affect soil formation as would cold temperatures and snow. The humus layer appears slightly hydromorphic due to the position of these profiles in depressions where runoff water accumulates.

Just as in Subalpine Rankers, the "rhizosphere" effect of the short vegetation combines with the insolubilizing potential of sesquioxides to rapidly immobilize near the surface the large amounts of soluble humic compounds produced by the vegetation.

REFERENCE: Carballas et al., 1967; Guitian Ojea and Carballas, 1968.

IV₃: HUMIC ANDOSOL (with little horizon differentiation)

(Andosol humique)

F.A.O.: *Humic Andosol*; U.S.: *Dystric Cryandept*

Location: Puy de Mercoeur, Puy-de-Dôme, France.
Topography: Moderate slope, northeast exposure. Elevation 1,150 m.
Parent material: Basalt pumice.
Climate: Atlantic montane. P. 1,300 mm; M.T. 6°C.
Vegetation: Beech forest with tall, hygrophilous, and nitratophilous grasses.

Profile Description

A_0A_1 (0-40 cm): Dark brown (7.5 YR 3/2) fibrous litter, grading into a crumb "mull"; less distinct crumb structure at depth, grading to pseudosilt with thixotropic (smeary) consistence; *diffuse* boundary.

$A_1(B)$ (40-75 cm): Dark brown (7.5 YR 3/3) highly humic pseudosilt; thixotropic, loose when dry; high water content at all times, very porous; many roots throughout.

B (75-80 cm): Thin rust band, becoming black (10 YR 2/1).

C: Very dark gray, slightly weathered pumice; often dry.

Geochemical and Biochemical Properties

Particle-size distribution: Difficult to define or measure due to the abundance of amorphous materials. In such typical Andosols, the content of clay minerals, *sensu stricto*, is low (a few percentage points).

Amorphous minerals and allophane: These consist of an aluminosilicate amorphous phase ("allophane" *lato sensu*) which is extractable using mild reagents. In a typical Andosol, they must represent at least 10% of the total mineral mass. Here, SiO_2 + Al_2O_3 equal about 20%; free iron alone represents 7%, or about 10% as Fe_2O_3. The total amorphous phase amounts to 30%.

Exchange complex: Exchange capacity is very high, but consists mainly of "pH dependent" charges (45 meq/100 g at pH 7 in A_1). S/T is always quite low in the Andosols of the Massif Central region (14% in A_1 and B). pH (water) is 4.5. Thus, this is an *oligotrophic Andosol* with acid mull.

Biochemical properties: Organic matter content is high, with values of 20% in A_1 and 11% in $A_1(B)$. C/N ratio of 13.5 at the surface and 12 in $A_1(B)$ is indicative of intense biological activity. Extractable carbon, using alkaline reagents at pH 12, is very high (75%). Fulvic and humic acids are very abundant because of insolubilization.

Physical properties: Important physical properties of Andosols to note are: low bulk density (less than 0.8 g/cm³); high porosity, permeability, and field capacity.

Genesis. Because of high porosity and high specific surface area, weathering proceeds rapidly. Vitreous elements become hydrated to form "allophane", which confers special properties to the soil. Allophane "stabilizes" soluble prehumus substances by making them insoluble and protecting them against further biodegradation through moderate polymerization (brown FA and HA). The very few complexes that migrate accumulate at the bottom of the moist zone, at the contact with the C horizon which remains drier (thin B horizon). Contrary to what is observed in Andosols with podzolic features, the majority of the free iron and aluminum resists translocation because of the abundance of these elements.

REFERENCE: Hetier, 1973.

IV₄: DIFFERENTIATED ANDOSOL [with (B) horizon]
(Andosol différencié)

F.A.O.: *Humic Andosol*; U.S.: *Dystric Cryandept*

Location: Le Bouchet, near Saint-Ours, Puy-de-Dôme, France.
Topography: Moderate slope. Elevation 900 m.
Parent material: Large blocks of pumiceous basalt.
Climate: Atlantic montane. P. 1,200 mm; M.T. 7°C.
Vegetation: Moorland with American bracken (*Pteridium aquilinum*) growing on former fallow land.

Profile Description

A_p (0-30 cm):   Dark brown (10 YR 3/3) humic layer; crumb structure; gradual boundary.
(B) (30-80 cm):   Yellowish-brown (10 YR 5/6); very fine crumb structure, thixotropic, loose when dry.
(B)C: Yellowish-brown weathered rock; porous.

Geochemical and Biochemical Properties

Particle-size distribution: Compared to the previous profile, the less intense development of this profile permits the identification of particle sizes. Clay content remains low (4%). Sand content is high (56%) and consists of unweathered rock fragments (slightly "vitreous").

Amorphous minerals and allophane: Although the (B) horizon is better differentiated, this profile is still a typical Andosol because the aluminosilicate amorphous material (allophane) in (B) is 14%. Free iron content is 3.6%.

Exchange complex: Much lower exchange capacity than in the previous soil with 20 meq/100 g in A_p and 9 meq/100 g in (B). On the other hand, base saturation is higher and increases from 14% at the surface to 48% at depth with a concomitant increase in pH from 5.2 to 6.4. Base saturation is underestimated in A_p because exchange capacity was measured at pH 7 and pH-dependent charges outnumber permanent charges.

Biochemical properties: Much less organic matter than in the previous profile [14.5% in A_p and only 2% in (B)]. C/N ratio is 12.5 in both A_p and (B) horizons.

Physical properties: Similar to those of previous profile but less pronounced.

Genesis. This profile is less weathered than the previous one because the basalt parent material is harder and formed of coarser elements (*vitric* tendency). The (B) horizon is very distinct from the A_p horizon. A slight tendency towards "brunification" is noticeable in the slightly humic (B) horizon. No doubt, this is due in part to a less montane type of climate. Contents of organic matter, which *decreases abruptly* from A_p to (B), and of "allophane" are lower. The differences in physical and chemical properties that have been mentioned for these two Andosols arise from the contrast in their ecology and evolution. These considerations justify the name "Differentiated Andosol" that Quantin (1974) has proposed for this type of soil.

REFERENCE: Hetier, 1973.

IV_5: ANDOPODZOLIC SOIL

(Sol andopodzolique)

F.A.O.: *Humic Andosol*; U.S.A.: *Dystric Cryandept*

Location: Summit of Puy de Dôme, Puy-de-Dôme, France.
Topography: Top of slope. Elevation 1,465 m.
Parent material: Projections of domite mixed with hornblende.
Climate: Upper montane. P. 1,300 mm; M.T. 4.5°C.
Vegetation: Pseudoalpine meadow.

Profile Description

A_0A_1 (0-20 cm): Very dark gray (10 YR 3/1) fibrous mor; changes imperceptibly into an A_1 with crumb structure; many roots.
B_{h1} (20-40 cm): Black (10 YR 2/1), very humic; massive structure; sticky and plastic when wet, loose when dry; gradual boundary.
B_{h2} (40-60 cm): Dark brown, less humic; same structure as above.
B_s (60-80 cm): Reddish-yellow (7.5 YR 7/8) loamy sand, containing domite fragments.
C: Grayish domite fragments.

Geochemical and Biochemical Properties

Particle-size distribution: Mineral material is loamy sand to sandy loam and contains practically no clay minerals.

Amorphous minerals and allophane: Not as rich in iron (2.4%) as the preceding soils due to the nature of the parent material. In B_h, 16% allophane-like compounds are found, which places this soil squarely within the Andosol class. However, migration of amorphous aluminosilicates, which represent only 4% of mineral matter in A_1, is substantial. The translocation index of these amorphous minerals is thus around 1/4 and slightly less for iron. This is a definite sign of podzolization.

Exchange complex: Exchange capacity is very high due to the abundance of organic matter (72 meq/100 g in A_1 and 52 meq/100 g in B_h). Base saturation is very low (from 3 to 4% depending on the horizon) which causes acidification and podzolization.

Organic matter: Very high organic matter content (33% in A, 27% in B_h, and still 9% in B_s). Substantial migration of the more mobile fulvic acids which are responsible for podzolization. However, the C/N ratio is low (about 13 in all horizons) as vegetation consists mainly of grasses.

Genesis. Podzolic features (with, if one is to judge by the thickness and aspect of the B_h horizon, a definite hydromorphic tendency) are superimposed on the features of an Andosol. Two reasons may be attributed to this evolution: (i) a very humid montane climate and (ii) the acidity of the parent material which is relatively low in iron, weatherable minerals, and calcium. The result is "climatic" podzolization. Despite the intense biological activity during the summer and a low C/N ratio, the production and massive migration of soluble organic compounds during certain periods of the year induce complexing and translocation of a part of the iron, but mostly of the aluminum. Silica which has been released migrates in solution. Thus, this soil displays the features of an Andosol and of a Podzol at the same time.

REFERENCE: Hetier, 1973.

IV₆: HYDROMORPHIC ANDOSOL

(Andosol hydromorphe)

F.A.O.: *Humic Andosol*; U.S.: *Aquic Dystrandept*

Location: Paramo de Gabriel Lopez, Cauca, Colombia.
Topography: Gentle slope. Elevation 3,170 m.
Parent material: Andesitic ash from the Purace Volcano.
Climate: Tropical montane. P. 2,000 mm; M.T. 9°C.
Vegetation: "Subparamo" forest (in the fog belt).

Profile Description

A_0 (10-0 cm):	Black (10 YR 2/1), very humic; fine blocky structure, thixotropic; very porous; many roots; clear boundary.
A_1 (0-50 cm):	Black (10 YR 2/1), very humic; coarse blocky to coarse prismatic structure, with vertical cracks; thixotropic, very porous; more and more plastic with depth; common roots.
A/C (50-65 cm):	Transition horizon; coarse blocky structure with dark organic coatings on peds; diffuse boundary.
C_g (65-130 cm):	Pale brown (10 YR 6/3); massive structure, very thixotropic.
C:	Partially weathered ash; less thixotropic than above.

Geochemical and Biochemical Properties

Amorphous minerals and allophane: From 28% in A_0, "allophane," *lato sensu*, decreases gradually throughout the humic horizons. In A/C, it still represents 17%, then remains constant at around 20% in C_g and C. The profile is low in clay minerals. Nevertheless, there is some metahalloysite in C_g.

Exchange complex: Exchange capacity is very high, above 120 meq/100 g in A_0 and A_1. On the other hand, base saturation is very low. The value of S barely exceeds 1 meq/100 g in both organic and mineral horizons (pH 4.7 in A_1).

Biochemical properties: Organic matter content is very high. It decreases from approximately 50% in A_0 and A_1, to 30% at the base of A_1, and to 15% in A/C. C/N ratio is 15.4 in A_1. The C_g horizon still contains 3.5% organic matter which contributes to its pale brown color.

Physical properties: Bulk density is very low at 0.55 g/cm³. Field capacity is considerable, but the water content at wilting point is also high at about 95% in A_0A_1.

Genesis. Under very humid equatorial climate, high altitude Andosols display extreme features. Humic horizons are very thick and consist almost entirely of organic and mineral amorphous materials (about 80% in A_0) with a great affinity for water. "Hydromorphism" is a constant characteristic in this type of profile. The organic horizons, which are almost constantly waterlogged, reveal an evolution towards peat as is evident from their coarse structure. Below these horizons, the partially bleached C_g horizon is infiltrated with the more mobile organic compounds, which partly reduce iron (early signs of evolution towards a *gley*). Finally, another peculiarity of equatorial Andosols developed from *volcanic ash* is noteworthy: i.e., the strong weathering of the C horizon [actually a (B)C] to a depth exceeding at times 1 m. Evidence of this is found in the high "allophane" content of the C horizon. It can be attributed to the importance of percolations, on the one hand, and to the high surface area of the finely divided material, on the other.

REFERENCE: Luna, C., and P. Faivre, Codazzi Institute, Bogota, Colombia, 1974.

PLATE IV

WEAKLY DIFFERENTIATED DESATURATED HUMIC SOILS

CHAPTER III

CALCIMAGNESIAN SOILS

Soils derived from parent materials containing limestone or dol-omite, with A_1C horizonation, or A(B)C in transitional forms. The exchange complex is saturated, or nearly saturated, with calcium and magnesium (except in some very humic forms).

"Rendzinas" which are the most typical Calcimagnesian soils contain "active carbonate" which slows the weathering process. It also immobilizes the abundant (in the more highly developed soils) and still incompletely transformed organic matter in a dark brown, or black, "rendzinic" horizon. This horizon has a coarse crumb structure. Aggregates are constituted by very stable humus-clay-carbonate complexes which often include some free iron.

Rendzinas are "intrazonal" soils (site climaxes). Consequently, their formation is tied to the calcareous parent material and is independent of climate. However, under humid climate, decarbonation may occur rapidly, especially in material containing silicate minerals, and lead to more or less intense "brunification." A whole series of intergrade soils can form between the Rendzina and the Brown soil, e.g., Brunified Rendzinas, Calcareous Brown soils, Calcic Brown soils. The degree of expression of the features peculiar to Rendzinas, especially the high organic matter content, decreases progressively within this series.

At high elevation, under humid or subalpine montane climate, decarbonation is accelerated, while a considerable amount of organic matter is added into the profile. Early immobilization of the organic matter is promoted by an edaphic factor (calcium) and a climatic factor (slower decomposition of fresh organic matter).

It is therefore possible to group soils developed from calcareous parent material into three subclasses with decreasing organic matter content.

Very humic Calcimagnesian soils (mountains)

The "Rendzina" feature of these soils is often lost because they are much higher in organic matter compared to mineral matter and most often depleted of carbonates. Effervescence is only observed in Humo-Calcareous soils, which are young Colluvial soils. Soils that are entirely depleted of "active" carbonate will be called "Humo-Calcic soils" or

"Humic Lithocalcic soils," depending on how much mineral impurities are contained in the humus and on the importance of the calcareous coarse fragments (Bottner, 1971).*

Humic Calcimagnesian soils, Rendzinas

These are characterized by the dark, thick humic horizon as defined above. Some forms, transitional with the next group, have a slightly developed, brown structural (B) horizon, which is often very stony (Brunified Rendzina).

Moderately humic Calcimagnesian soils

In this group, forming a transition towards Brunified soils, decarbonation is more intense, at least in the surface horizon. This leads to a thinner humic A_1 horizon, generally lighter in color compared to the brown and well-developed normal structural (B) horizon. These soils have formed from mixed parent materials with a lower proportion of carbonates relative to silicate minerals. Decarbonation may be rapid. In *Calcareous Brown soils*, the (B) horizon still contains some active carbonate, whereas the (B) horizon of *Calcic Brown soils* is totally decarbonated. Effervescence is observed only at depth in a (B)C or C_{ca} horizon formed by the reprecipitation of leached carbonates. Finally, if a profile is devoid of active carbonates in all horizons (e.g., a soil overlying a hard limestone slab), it should not be classified with the Calcimagnesian soils (see Profile IX$_4$, Eutrophic Brown soil, Chapter V).

Genesis. The genesis of Calcimagnesian soils is linked to the rate of depletion of active carbonates from the whole profile and is thus highly dependent on the nature of the parent material.

In plains, four types of parent materials will be considered. Decarbonation is very slow, or does not occur at all, on very pure and soft limestone (like chalk, for instance) or on regularly renewed "colluvium" (Table 3). On such materials, Rendzinas reach a stable equilibrium. On the contrary, carbonates are rapidly lost from marl or marly limestone. A fairly abundant silicate phase subsists and forms the (B) horizon of "Calcic Brown soils." The typical Rendzina, with an A_1C profile, is seldom observed on marly limestone, because a more or less well-developed brunified (B) horizon forms very rapidly. The true Rendzina can only develop on more or less eroded slopes.

The genesis of soils derived from a more or less eroded, or reworked, terra fusca layer, overlying hard limestone, is peculiar. Because of the presence of terra fusca, the soil profile immediately acquires the features of a "Brown soil." This Brown soil is usually "calcic" with an effervescent C_{ca} horizon. In the absence of the C_{ca} horizon, it will be "eutrophic" or even "mesotrophic." This occurs when the terra fusca rests directly on a

---

*The former Carbonated Humic soil designation has not been retained because of a possible confusion. Such soils, although rich in calcareous coarse fragments, are often low in active carbonate.

limestone slab or when it is mixed with very hard and slowly soluble limestone fragments (Profile IX$_4$, Chapter V).

The Rendzina phase appears only after *secondary recarbonation* of the whole profile. Recarbonation generally results from a mechanical mixing of terra fusca with limestone fragments through erosion or cryoturbation. Such a soil is seldom a typical Rendzina but rather a "secondary" Brunified Rendzina. Due to its "polycyclic" evolution, this type of soil is quite different from the primary (or monocyclic) Brunified Rendzina. The latter soil results from direct genesis whereby humus is incorporated into the weathered products of marl or marly limestone.

Note 1. In the chapter on Fersialitic soils (Chapter VIII), I will mention the existence of "secondary Red Rendzinas" and "Calcareous Brown soils" which have developed from similar rubefied fersialitic parent material. The latter soils are more frequent in regions with warm climate, where "terra rossa," which is a Rubefied Paleosol, replaces "terra fusca" on hard limestone.

Note 2. Evolution towards brunification is especially rapid on some rock outcrops such as arenaceous and dolomitic limestones. Here, the typical crumb structure of Rendzinas is less evident, consistence is loose, texture is sandy, and the humus frequently evolves towards a moder. These soils are called Pararendzinas by German authors (Profile VI$_2$).

Table 3.   Genesis of Calci-Magnesian Soils

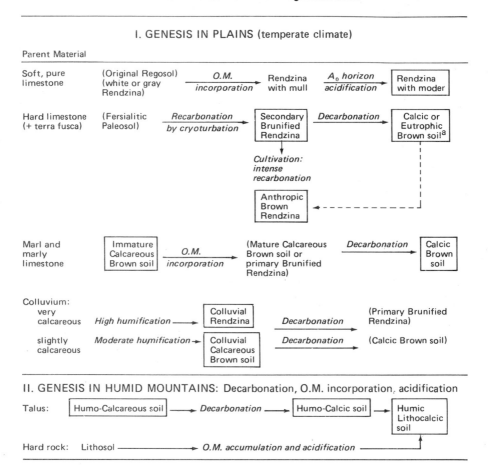

I. GENESIS IN PLAINS (temperate climate)

Parent Material

Soft, pure limestone → (Original Regosol) (white or gray Rendzina) — *O.M. incorporation* → Rendzina with mull — *A₀ horizon acidification* → | Rendzina with moder |

Hard limestone (+ terra fusca) → (Fersialitic Paleosol) — *Recarbonation by cryoturbation* → | Secondary Brunified Rendzina | — *Decarbonation* → | Calcic or Eutrophic Brown soil[a] |

↓ *Cultivation: intense recarbonation*

| Anthropic Brown Rendzina | ← - - - - - - - - - - ┘

Marl and marly limestone → | Immature Calcareous Brown soil | — *O.M. incorporation* → (Mature Calcareous Brown soil or primary Brunified Rendzina) — *Decarbonation* → | Calcic Brown soil |

Colluvium:
very calcareous — *High humification* → | Colluvial Rendzina | — *Decarbonation* → (Primary Brunified Rendzina)

slightly calcareous — *Moderate humification* → | Colluvial Calcareous Brown soil | — *Decarbonation* → (Calcic Brown soil)

II. GENESIS IN HUMID MOUNTAINS: Decarbonation, O.M. incorporation, acidification

Talus: | Humo-Calcareous soil | —→ *Decarbonation* —→ | Humo-Calcic soil | → | Humic Lithocalcic soil |

Hard rock: Lithosol ——————→ *O.M. accumulation and acidification* ——————

[a]See Profile IX₄, Eutrophic Brown soil on terra fusca (Chapter V).

V₁: COLLUVIAL RENDZINA (with mull)

(Rendzine colluviale)

F.A.O.: *Rendzina*; U.S.: *Typic Rendoll*

Location: Bellefontaine Dell, near Nancy, Meurthe-et-Moselle, France.
Topography: Dry dell at foot of slope, southwest exposure. Elevation 260 m.
Parent material: Slope colluvium consisting of Bajocian limestone fragments, terra fusca, and loam.
Climate: Atlantic with continental trend. P. 740 mm; M.T. 9.4°C.
Vegetation: Beech-grove with hornbeam, common maple and calcicole shrubs (*Viburnum lantana, Cornus sanguinea*); flora of eutrophic mull consists of *Asarum europaeum, Anemone hepatica, Asperula odorata, Euphorbia sylvatica,* etc.

Profile Description

A₁ (0-15 cm):   Gray-brown (7.5 YR 5/1) very humic mull; fine (0.2 cm) crumb structure; many angular limestone pebbles; very porous; many roots; diffuse boundary.

A₁C (15-40 cm):   Brown (7.5 YR 5/2); very humic; crumb structure; well aerated; more limestone gravel than above; common roots; clear boundary.

C: Bajocian limestone talus embedded in an ocherous matrix of fine particles.

Geochemical and Biochemical Properties

Particle-size distribution: Cannot be precisely determined because of the high organic matter content. Texture is well graded. Sand particles consist of limestone fragments. Relatively high contents of silt and clay (20% clay), deriving from terra fusca and reworked loam.

Exchange complex and carbonates: Very high exchange capacity due to the high organic matter content (50 meq/100 g in A₁ and 30 meq/100 g in A₁C). CaCO₃ content (as coarse fragments) is high and increases from 17% in A₁ to 37% in A₁C. "Active" carbonate shows less variation and ranges from 10 to 12%. The exchange complex is saturated, mainly with calcium and magnesium (Mg^{2+} exceeds 10 meq/100 g in the surface horizon). pH (water) is 7.5 in the surface horizon.

Biochemical properties: Exceptionally high organic matter content. It decreases rapidly from more than 20% in the surface horizon to 8% in A₁C. C/N ratios of 13 in A₁ and 10 in A₁C indicate intense biological activity. Extractable carbon is very low (17%), even after decarbonation. The FA/HA ratio is considerably larger than 1.

Free iron: Iron is inherited from the terra fusca and reaches 0.75% in A₁C, which is high for a Rendzina.

Genesis. This is a typical, very humic forest Rendzina, in which renewal of carbonates is ensured by the intense mixing of the colluvial materials. Mixing is both mechanical (deposition) and biological (animal activity). As a large proportion of the material is inherited (iron and clay from the terra fusca), this soil profile could be designated "polycyclic." The abundant active carbonates block humification at an early stage by slowing the mineralization of the incompletely transformed humic compounds. The content of nonextractable humin is characteristically high. This humin, either "microbial" or "inherited," is young. The weak maturation of the humus contrasts with the strong maturation of the humus in Chernozems. Very active nitrogen fixation and nitrification are evidences of intense biological activity, which because of a high rate of CO_2 production is the cause of the slight decarbonation at the surface.

REFERENCE: Dommergues and Duchaufour, 1966: *Soil R.C.P. 40.*

V₂: HUMIC RENDZINA WITH MODER

(Rendzine humifère à moder)

F.A.O: *Rendzina*; U.S.: *Typic Rendoll*

Location: Liffol-le-Grand Forest, Vosges, France.
Topography: Plateau, very gentle slope. Elevation 420 m.
Parent material: Oolithic Rauracian limestone.
Climate: Montane trend. P. 1,000 mm; M.T. about 9°C.
Vegetation: Beech forest with common oak, hornbeam and Norway maple, in lesser amounts; sparse herbaceous plants, moss.

Profile Description

LA_0 (5-0 cm):   Dark reddish-brown moder, fibrous organic horizon; many roots.
A_1 (0-30 cm):   Dark brown (10 YR 3/3); coarse (8 mm) crumb structure; very high porosity; many roots; 20% pebbles and gravel increasing to 35% at the base; strong biological activity; abrupt boundary.
C: Disintegrating oolithic limestone; embedded in white (10 YR 8/2) finely divided chalk, slightly hard when dry, sticky when wet; few coarse roots along cracks.

Geochemical and Biochemical Properties

Particle-size distribution: Can be interpreted only in relation to the total carbonate content which represents more than 99% of the C horizon and is responsible for its sandy texture. However, carbonate content is only 7% at the top of A_1. A clay content of 11% in this horizon illustrates the concentration of silicate impurities due to decarbonation.

Exchange complex: Except for the desaturated A_0 horizon (pH 5.2), S/T is 100%. Exchange capacity is about 28 meq/100 g.

Biochemistry: Organic matter is not as well incorporated as in the Colluvial Rendzina. Its content is high and decreases from 15% at the top of A_1 to 4.8% at its bottom. Only traces are found in the C horizon. C/N ratio drops abruptly from 32 in A_0 to 14 at the top of A_1 and 11 at its base. Extractable carbon is very low throughout.

Free iron: Iron is concentrated, even more than clay, in the upper part of A_1 as a "residual" element of decarbonation. From 1.5% just below the A_0 horizon, it decreases to 0.6% at the 20-cm depth and to 0.02% in C. The biogeochemical cycle (input from litter) plays a nonnegligible role in this surface concentration.

Genesis. The genesis of this Rendzina with moder, formed on undisturbed limestone, clearly differs from that of the Colluvial Rendzina. In the present profile, mixing of organic matter and calcareous material does not result from mechanical action but solely from biological activity. Thus, mixing is less complete. This explains the formation, at the surface, of an A_0 horizon with slow decomposition, which is further enhanced by a montane type of climate. The very fast and marked biological decarbonation, which is governed by the acid forest humus, results in the high concentration of clay and iron at the surface. Because the content of silicate particles is insufficient to promote brunification, this soil remains a Rendzina. Forests are rarely found on soils with parent material that weathers to "chalk"; they are usually covered with *shrubs* or meadow (less humic Gray Rendzinas). But in this case, the presence of cracks in the parent material has enabled deep-rooted trees to grow.

REFERENCE: Le Tacon, F., Forest Soils Laboratory, C.N.R.F., Amance, 1973.

V₃: HUMO-CALCAREOUS SOIL WITH MULL

(Sol humo-calcaire à mull)

F.A.O.: *Rendzina*; U.S.: *Cryic Rendoll*

Location: Noirvaux Dell, Amancey, Doubs, France.
Topography: Steep slope at foot of a Rauracian calcareous cliff, northeast exposure. Elevation 500 m.
Parent material: Upper Jurassic coarse calcareous talus, nonstabilized.
Climate: Montane (emphasized by topography). P. about 1,500 mm; M.T. about 7°C.
Vegetation: Maple (*Acer pseudoplatanus* and *A. campestre*), *Fraxinus excelsior*, *Phyllitis scolopendrium*, *Aspidium aculeatum*.

Profile Description

A_{11} (0-30 cm): Very dark brown (7.5 YR 2/2) very humic calcareous mull; irregular and well-aerated coarse crumbs; much calcareous gravel and cobbles (90%); gradual boundary.

A_{12} (30-60 cm): Gravel and cobbles amount to 95-98% of bulk mass; small amount of very dark brown (7.5 YR 2/2) fine earth, dispersed among the coarse fragments, consisting of same type of humic mull as above; crumbs are more angular.

A_1C: Lower in humus; amount of fine earth decreases progressively with depth.

Geochemical and Biochemical Properties

Particle-size distribution: Cannot be precisely estimated in this very humic and very calcareous material. The silicate fraction of the fine earth is very clayey (about 50% clay) and consists of reworked terra fusca. Calcium carbonate represents 30% of the fine earth (11% active) in A_{11} and 45% (16% active) in A_{12}.

Biochemistry: Humus consists of a calcareous mull exceptionally high in organic matter (29% in A_{11}, 26% in A_{12}, and still 15% of A_1C). A characteristic of these soils is a slow decrease in organic matter content with depth. C/N ratio ranges between 15 and 16 depending on the horizon.

Exchange complex: Very high exchange capacity related to the high content of humified organic matter. Exchange complex is base-saturated (pH 8.1).

Genesis. Formerly designated "Carbonated Humic soil," this soil has now been called "Humo-Calcareous soil" by Bottner (1971). It is characteristic of *nonstabilized* talus, in mountains with cold exposure. It contrasts with Humo-Calcic soils formed on stabilized talus. These are decarbonated and acidified at the surface, but occur under similar topographical and climatic conditions. The supply of calcareous material is constant, as indicated by the very high content of calcareous coarse fragments, the incorporation of organic matter to great depth, and the high content of active carbonates. The vegetation, consisting mainly of maple trees and ferns (*Scolopendrium*), is well adapted to this particularly unstable soil, yet rich in plant-available minerals and nitrogen due to its biochemical properties. Because of a low content in silicate minerals, the humus forms a special kind of "calcareous mull," called "mull-like moder" by Kubiena. Once stabilized, this type of talus material becomes covered with acidifying evergreens and Ericaceae, which produce a "tangel-like," or even "mor," humus at the surface. Decarbonation then proceeds very rapidly (see Profile V₄).

REFERENCE: Gaiffe (Mrs.), unpublished report from C.N.R.S. Nancy and Soil Survey Laboratory, University of Besançon. Photo by M. Gury.

V₄: HUMO-CALCIC SOIL WITH TANGEL

(Sol humo-calcique à tangel)

F.A.O.: *Humic Cambisol*; U.S.: *Entic Cryumbrept*

Location: Grande-Chartreuse Massif, Isère, France.
Topography: Talus at the foot of Urgonian cliff, northwest exposure. Elevation 1,200 m.
Parent material: Talus of Urgonian limestone.
Climate: Exceptionally cold soil climate, considering the elevation, due to both site and exposure factors; permafrost under the unconsolidated material. M.T. about 5°C.
Vegetation: Subalpine moorland with *Rhododendron ferrugineum*, *Vaccinium vitis-idaea*, *Arctostaphylos alpina*, moss, etc.

Profile Description

A$_{00}$ (15-0 cm):	Dark brownish-red fibrous mor; many Ericaceae roots.
A$_0$H (0-30 cm):	Black (10 YR 2/1) "tangel"; very fine crumb structure; sticky consistence, very porous; presence of small fecal pellets; many fine roots; skeletal limestone cobbles; weakly effervescent at places.
A$_1$ (30-80 cm):	Black (10 YR 2/1) mull-moder; fine to medium crumb structure; weakly effervescent; unsorted angular limestone gravel with common spots of more or less sticky humus; skeletal arenaceous limestone.
C:	Calcareous talus.

Biochemical Properties

Almost pure and very slightly transformed organic matter at the surface, becoming progressively richer in mineral particles, and more humified but less acid with depth. Organic matter content is still 56% in A$_0$H and 18% in A$_1$. C/N ratio is high (35 in A$_{00}$) but decreases to 20 in A$_1$. Traces of inactive carbonates are found at the level of the "tangel" horizon and 1% in the mull-moder horizon. *Base saturation* increases from 33% in A$_{00}$ (pH in water 4.5) to 80% in A$_0$H (pH 5.9). The A$_1$ horizon (pH 7.2) is almost base-saturated. The *exchange capacity* of the tangel horizon (120 meq/100 g) is twice that of the mor in the A$_{00}$ horizon (60 meq/100 g), which correlates well with the high degree of humification in A$_0$H.

Genesis. In contrast with Humo-Calcareous soils, which typically form on nonstabilized talus, Humo-Calcic soils (Bottner, 1971) are found on stabilized talus with cold exposure. Such conditions favor the accumulation of increasingly acid organic matter which, at least in the surface horizon, decays more slowly. A calcifugous *subalpine moorland* replaces the mull flora and favors the formation of a mor. At the same time, decarbonation proceeds to greater depth under the influence of the soluble organic compounds. These form at the surface, then infiltrate and precipitate in calcium-rich horizons, while active carbonates are eliminated as soluble Ca(CO$_3$H)$_2$ (Bottner, 1971). A succession of three types of humus is thus observed: a fibrous acid mor at the surface, a more humified and less acid tangel in the middle, and a calcium-saturated "mull-moder" low in clay at the base of the profile. The exceptionally cold soil climate at this site has favored this evolution as indicated by the subalpine flora which normally occurs at the montane elevation zone.

REFERENCE: Bartoli, 1962.

V₅: HUMIC LITHOCALCIC SOIL WITH HYDROMODER

(Sol lithocalcique humifère à hydromoder)

F.A.O.: *Rendzina*; **U.S.:** *Cryic Rendoll*

Location: Raxplateau, Austrian Alps, Austria.
Topography: Platform with gentle northwest slope. Elevation 1,950 m.
Parent material: Upper Jurassic white "Wetterstein" limestone and terra fusca.
Climate: Alpine elevation zone; high snowfall; cold and wet soil climate.
Vegetation: Alpine meadow with *Sesleria varia, Carex firma, Dryas octopetala, Loiseleuria procumbens, Silene acaulis.*

Profile Description

A_0 (0-8 cm): Black (10 YR 2/1) very humic, sticky hydromoder; very fine coprogenic crumbs; many roots.
A_1 (8-20 cm): Very dark grayish-brown (10 YR 3/2), very humic; homogeneous massive structure, breaks into irregular blocks; many roots.
A_1C (20-35 cm): Dark brown (10 YR 3/3) humus, mixed with terra fusca and calcareous gravel; compact and homogeneous; weakly effervescent.

Biochemical Properties

The slightly acid and noncalcareous organic matter is intimately mixed with very fine loam (56% organic matter, 43% mineral matter). The organic matter content drops to 28% in the terra fusca-enriched A_1C horizon. *Base saturation* increases from 72% in the upper horizons to 100% in the slightly calcareous A_1C horizon. The exchange capacity of the organic matter is about 200 meq/100 g. The high level of base saturation is maintained to a great extent by the upward movement of limestone gravel through cryoturbation. This process does not occur in neighboring profiles formed on unfragmented limestone slab; the humus is then very acid (pH in water 5) and strongly desaturated. The loam mixed with the surface organic matter does not derive from the limestone, as indicated by its mineralogical composition. It contains quartz, mica, and orthose which demonstrates an eolian origin (Solar, 1964).

Genesis. This soil corresponds to the "Pechrendzina" described by Kubiena (1953) and Solar (1964). This is a broader concept of the Rendzina, defined as a "humic soil with AC horizonation formed on limestone." But I would prefer to restrict the concept of Rendzina to soils with well-separated crumbs consisting of a mixture of humus, clay, and carbonates. This is not the case here. As the organic matter content far exceeds the mineral matter content, I prefer the designation "Humic Lithocalcic soil with hydromoder," as proposed by Bottner (1971). An alpine meadow and an almost permanently wet moder replaces the acid mor described in the previous soil. The moder is also biologically more active and much higher in mineral matter than the mor. A fraction of the mineral impurities is of eolian origin while the rest (in A_1C) derives from terra fusca. Differences between the two types of humus are related to vegetation. The alpine meadow provides a more favorable environment for biological activity than does the moorland with Ericaceae.

REFERENCE: Solar, 1964.

V$_6$: HUMIC LITHOCALCIC SOIL WITH MOR
(Sol lithocalcique humifère à mor)

F.A.O.: *Humic Cambisol*; U.S.: *Lithic Cryumbrept*

Location: Moucherotte, Vercors Massif, Isère, France.
Topography: 20° slope, northwest exposure. Elevation 1,900 m.
Parent material: Eroded Urgonian limestone (*lapiaz*).*
Climate: Humid subalpine. P. 1,250 mm; M.T. 6.2°C at Villard-de-Lans.
Vegetation: *Lycopodio-mugetum, Pinus montana uncinata* and Ericaceae (*Rhododendron ferrugineum, Vaccinium myrtillus, V. uliginosum, V. vitis-idaea, Empetrum hermaphroditum*, etc.).

Profile Description

A$_{00}$ (0-5 cm):   Very dark brown, fibrous mor with 60% identifiable plant debris; spongy structure; mycelium filaments.

A$_0$H (5-30 cm):   Dark reddish-brown (5 YR 2/2) humus, more decomposed than above, 80% of finely divided and nonidentifiable plant debris; fine, fairly compact structure; many roots.

A$_1$ (30-33 cm):   Black (5 YR 2/1) mull-moder forming a thin band at the contact with the limestone and filling the *lapiaz* cracks; crumb structure.

R: Limestone slab, eroded into *lapiaz*.

Biochemical Properties

Except in the weakly developed and often absent A$_1$ horizon, the whole profile is formed of a very acid "mor," with very few noncalcareous mineral impurities (6-8% in A$_0$H). Silicate minerals content is 35% in A$_1$. Properties are indeed those of a mor, not those of a tangel: high C/N ratio (about 40 in A$_{00}$, 29-30 in A$_0$H); pH is 4; S/T is about 20% in A$_0$H.

Biochemical properties change drastically in the A$_1$ horizon with a C/N ratio of 22, a pH of 6.8, and S/T of 70%.

Exchange capacity, which is a reflection of humification, changes only slightly throughout most of the profile. It equals 200 meq/100 g of organic matter but jumps to 400 meq/100 g in A$_1$. Similarly, extractable carbon by Na-pyrophosphate is less than 10% in A$_0$H, but increases abruptly to 45% in A$_1$.

Physical properties and soil climate have been investigated by Gilot and Dommergues (1967). Bulk density is very low (0.14 g/cm^3). The humus becomes water-saturated at snowmelt and remains cool and wet throughout summer.

Genesis. This is an organic soil consisting almost exclusively of an acid mor. It is very close to a "hydromor" due to the characteristics of its soil climate. This causes the formation of an extensive, but weakly humified A$_0$H "fine mor." It may be asked why such an unfavorable humus—not even a "tangel" according to Kubiena—has formed on limestone. The underlying, undivided, very hard, and insoluble limestone plays no part because it is inactive. Its mixing with organic matter is not assured and is, in fact, restricted to the thin A$_1$ horizon. Other environmental factors (such as elevation, very acidifying vegetation, and exposure) are adverse to biological activity. It is possible, however, that mountain pine, which is indifferent to soil reaction, established itself at this site by using the *lapiaz* cracks. The thin mor layer that developed then allowed the calcifugous Ericaceae to grow and build a thick layer of very acid, raw humus.

REFERENCE: Gilot and Dommergues, 1967.

*Translators' note: The word *lapiaz* is a local term used in the Jura region to designate the superficial channelling of limestone by runoff water.

PLATE V

HUMIC AND VERY HUMIC CALCIMAGNESIAN SOILS

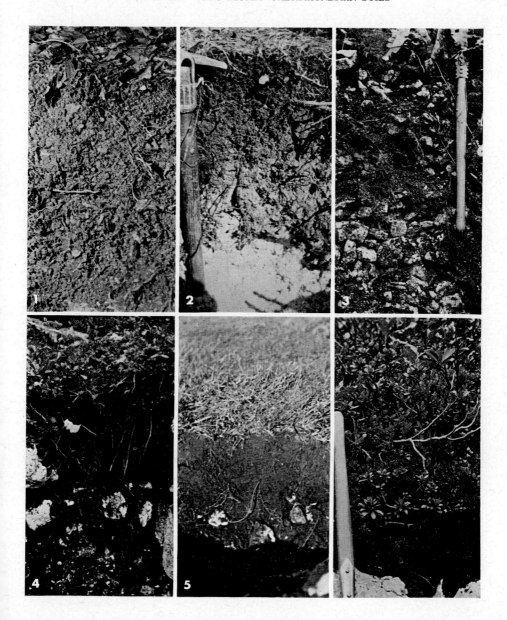

VI₁: SECONDARY (polycyclic) BRUNIFIED RENDZINA

(Rendzine brunifiée secondaire)

F.A.O.: *Rendzina*; U.S.: *Eutrochreptic Rendoll*

Location: Saint-Fiacre Wood, Blenod-lès-Toul, Meurthe-et-Moselle, France.
Topography: Edge of plateau (Côtes de Meuse), very gently sloping, south exposure. Elevation 375 m.
Parent material: Gravelly Argovian limestone and reworked terra fusca.
Climate: Atlantic with continental trend. P. 800 mm; M.T. 9.4°C.
Vegetation: Beech grove with sycamore maple; neutrophilous and xerothermophilous vegetation consisting of *Cornus mas, Cornus sanguinea, Sesleria coerulea, Hepatica triloba*.

Profile Description

A_1 (0-20 cm):	Dark reddish-brown (5 YR 2/2), very humic, calcareous mull, loamy with few coarse fragments at the surface, then more clayey with coarse crumbs and much gravel; very porous; coarse horizontal roots; strongly effervescent; gradual boundary.
(B) (20-40 cm):	Yellowish-red (5 YR 4/6) silty clay loam; much unsorted gravel; medium blocky structure, very porous; few coarse diagonal and horizontal roots; abrupt wavy boundary.
C_{ca}:	Hard calcareous crust; limestone gravel embedded in white, precipitated calcareous tufa.

Geochemical and Biochemical Properties

Particle-size distribution and free iron: Texture indicates that the parent material is a reworked terra fusca mixed with loam. Clay content is 16% in A_1 and 31% in (B). This is further indicated by a much lower iron content (1.2%) than in pure terra fusca.

Exchange capacity and carbonates: Relatively low exchange capacity, considering the amounts of clay and organic matter present [20 meq/100 g in A_1 and 12 meq/100 g in (B)]. This is due to the large proportion of kaolinite in the clay minerals derived from terra fusca. Large amounts of $CaCO_3$, with 41% in A_1 and 58% in (B), of which 8.6 and 11% is "active" $CaCO_3$, respectively. Under these conditions, the exchange complex is saturated.

Biochemistry: Organic matter is abundant, as is usually the case in the presence of active $CaCO_3$ [8% in A_1, but only 2.8% in (B)]. C/N ratio of 16 in A_1.

Genesis. This polycyclic soil is formed on old parent material, which is a true Fersialitic Paleosol containing terra fusca and reworked loam. The secondary evolution towards a Rendzina, with high incorporation of humus and early blocking of humification by carbonates, has been promoted by special ecological conditions: (i) rather soft granular limestone, fractured and weathered by cryoturbation and other mechanical agents; (ii) topographical position favoring lateral translocation (mechanical or chemical) of fine soil into A_1 and (B) and of $CaCO_3$ by dissolution and reprecipitation at the break of slope. This latter process is responsible for the formation of a crust which is normally replaced by a simple pseudomycelium (see Profile VI₃). Under these conditions, the "acidifying" effect of the forest humus has become entirely inoperative.

REFERENCE: *Soil Map of France (1:100,000)*, Nancy sheet.

VI$_2$: DOLOMITIC PARARENDZINA
(Pararendzine dolomitique)

F.A.O.: *Calcic Cambisol*; U.S.: *Lithic Eutrochrept*

Location: Mount Parnassus, Greece.
Topography: Valley side, 35% slope. Elevation 1,320 m.
Parent material: White dolomite.
Climate: Mediterranean montane with very dry summer.
Vegetation: Pine forest with black pine *(Pinus laricio austriaca)*.

Profile Description

A$_0$ (3-0 cm):   Organic xeromoder, fibrous and granular.
A$_1$ (0-15 cm):   Dark reddish-brown (5 YR 3/4) sandy loam; fine crumb structure; *slowly* effervescent; very porous; many medium and fine roots; clear smooth boundary.
C (15-50 cm):   White weathered dolomite, sandy loam texture; massive, loose; gradual boundary with compact dolomite blocks.

Geochemical and Biochemical Properties

Particle-size distribution and geochemistry: Fine sandy loam texture with 30% calcium carbonate. The parent material consists of almost pure dolomite (2.2% silica, traces of alumina) and is practically devoid of clay. Clay minerals (mainly illite and chlorite) increase to 12% at the surface. This indicates strong decarbonation with concentration of silicate impurities. For the same reason, free iron increases from 0.1% in C to 1.7% in A$_1$.

Exchange complex: Saturated mainly with calcium and magnesium (pH in water is 8).

Biochemistry: High organic matter content in A$_1$ (14%) with a C/N ratio of 23, indicating a relatively low level of biological activity. This is supported by three factors: (i) a very dry soil climate, (ii) the nature of the debris (pine needles), and (iii) the lower efficiency of dolomitic with respect to calcareous limestone in promoting biological activity.

Genesis. As in the Rendzina with moder under humid climate, the genesis of this Rendzina proceeds slowly because of the exceptionally high content of carbonates which prevents intense brunification. However, two ecological differences are of importance: (i) the very dry soil climate during the summer which slows the humification process; and (ii) the dolomitic, thus magnesian, nature of the parent material. Dolomite is less efficient than "active" calcium carbonate in promoting humification. The crumb structure is finer and less clearly expressed. A true granular moder forms at the surface. Dolomite grains, which are relatively inert and difficultly attacked by CO_2, confer to the soil a sandy loam texture that is comparable to that of Brunified Rendzinas formed on arenaceous limestone. These soils were called "Pararendzinas" by German authors. Furthermore, iron oxides are not completely incorporated within the clay-humus aggregates of the soil surface horizon. They are better separated and somewhat rubefied under the influence of the climate and provide the soil with its reddish-brown color.

REFERENCE: Desaunettes, 1970.

VI₃: ANTHROPIC BROWN RENDZINA

(Rendzine brune anthropique)

F.A.O.: *Rendzina;* U.S.: *Lithic Rendoll*

Location: Villey-le-Sec, Meurthe-et-Moselle, France.
Topography: Plateau sloping gently towards the West. Elevation 290 m.
Parent material: Bajocian oolithic limestone and terra fusca.
Climate: Atlantic with continental trend. P. 730 mm; M.T. 9.5°C.
Vegetation: Xerophilous meadow consisting of *Brachypodium pinnatum*, *Bromus erectus*, *Festuca duriuscula*, *Hippocrepis comosa*, and *Coronilla varia*.

Profile Description

A_p (0-22 cm)   Dark brown to brown (7.5 YR 4/3) calcareous humic mull; strongly effervescent; well-separated rounded crumbs; well aerated; much calcareous gravel; many fine roots; clear boundary.
C (22-40 cm):   Platy oolithic limestone, penetrated at places by terra fusca.
C_{ca}:   White, powdery "pseudomycelium" of precipitated carbonates between calcareous coarse fragments.

Geochemical and Biochemical Properties

Particle-size distribution and carbonates: Very high content of carbonates (56%) in the fine earth, of which 7.3% is "active." The silicate fraction is very rich in clay and sesquioxides (55% clay, 7% free iron, and 0.8% free alumina). These are characteristic values of terra fusca.

Exchange complex: The exchange capacity is moderate (14 meq/100 g) because of the high content of carbonates, but it is high on a silicate plus humus basis. With a pH (water) of 7.5 in A_1, the exchange complex is definitely saturated.

Biochemistry: The organic matter content (5%) is lower than in the other Rendzinas and not sufficient to mask the color of iron. C/N ratio of 10.9 is typical of fallow xerophilous meadows.

Genesis. This profile can be identified as a former secondary Brunified Rendzina. It was formed under forest cover and was thus very humic at first, but was subsequently transformed through cultivation. The composition of the silicate fraction indicates that it is "terra fusca," or clay derived from previous decarbonation. The formation of a pseudomycelium in C_{ca} can be attributed to the initial forested stage. Cultivation is at the origin of a double process. It has caused fragmentation of the coarse fragments with a considerable increase in carbonates in the fine earth and dehumification of the most labile fraction of the organic matter. Consequently, the brown color due to iron, which was partly masked in the humic "rendzinic" horizon initially, could develop. Such an evolution is comparable to that of the "Red Rendzinas" of southwestern France. If profiles developed under natural conditions must be taken as the basis for classification and nomenclature purposes, this soil may be called an *Anthropic Brown Rendzina*.

REFERENCE: Gury, M., *Legend, Soil Map of the Plateau of Haye,* C.N.R.S., Nancy, 1972.

VI₄: VERTIC CALCAREOUS BROWN SOIL

(Sol brun calcaire vertique)

F.A.O.: *Vertic Cambisol;* U.S.: *Vertic Eutrochrept*

Location: Guadalquivir River basin, near Cordova, Spain.
Topography: Gentle slope. Elevation 90 m.
Parent material: Mio-Pliocene marl.
Climate: Mediterranean, dry summer season. P. 500 mm; M.T. 19.6°C at Sevilla.
Vegetation: Cultivated.

Profile Description

A_p (0-20 cm):   Olive-yellow (5 Y 6/3), slightly humic clay; medium crumb structure; slightly porous; fine roots; clear wavy boundary.

(B) (20-50 cm):   Olive-yellow (5 Y 6/3) clay; subangular blocky to prismatic structure; hard when dry, slightly porous; diffuse wavy boundary.

$(B)_g$ (50-100 cm):   Olive-yellow with large diffuse and irregular yellowish-brown (10 YR 6/6) or white spots; strong prismatic structure with wide cracks; very hard when dry; diffuse boundary.

C: Olive-brown marl.

Geochemical and Biochemical Properties

Particle-size distribution and carbonates: Clay texture with 45% clay and 31% $CaCO_3$ is fairly constant throughout the profile. The major clay mineral is montmorillonite with illite and chlorite as accessory minerals.

Exchange complex: Exchange capacity of 19 meq/100 g, saturated with calcium and magnesium (2-2.4 meq Mg^{2+}/100 g).

Biochemistry: Very little organic matter (1%), evenly distributed within the A_p horizon. C/N ratio is about 9.

Iron: Much less (0.7%) than in the Colluvial Calcareous Brown Soil with reworked terra fusca (Profile VI₅).

Genesis. This soil is immature, as it has formed on a soft parent material subject to strong erosion. The absence of decarbonation and the constancy of the $CaCO_3$ content throughout the profile underline the young age of this soil. Most of the mineral constituents (clays, carbonates) are inherited from the parent material without major transformation. This also explains the low organic matter content, as the humic horizons have been truncated before carbonates had time to influence the humification process. This type of soil is very close to a Regosol and is frequently observed on marl or marly limestone under Mediterranean climate. Such a soil is not easily classified. Several features bring it close to a Vertisol, such as the abundance of swelling minerals, desiccation cracks during the dry season, etc., but it is less developed, especially with regard to organic matter. Furthermore, "mottling," i.e., the presence of gray or ocherous spots, is a sign of moderate hydromorphism, which causes some iron to segregate. Without erosion, such a soil would evolve towards a more humic and partly decarbonated Calcareous Brown soil (with a ca horizon), or even towards a "monocyclic" Calcic Brown soil.

REFERENCE: Albareda, J.M., *Guidebook, Conference on Mediterranean Soils*, Madrid, 1966.

VI₅: COLLUVIAL CALCAREOUS BROWN SOIL
(Sol brun calcaire colluvial)

F.A.O.: *Calcic Cambisol;* U.S.: *Mollic Eutrochrept*

Location: Sexey Wood, Haye Forest, Meurthe-et-Moselle, France.
Topography: Dry valley floor, carved in the Bajocian plateau. Elevation 215 m.
Parent material: Fine loam colluvium, terra fusca and some calcareous gravel.
Climate: Atlantic with continental trend. P. 740 mm; M.T. 9.4°C.
Vegetation: Forest of common oak, hornbeam, and ash with many calcicole shrubs. Flora of *active mull* is neutrophilous *(Arum maculatum, Asarum europaeum).*

Profile Description

A₁ (0-10 cm):	Yellowish-brown (10 YR 5/4) mull, becoming lighter with depth; crumb structure, well aerated; many roots; gradual boundary.
(B) (10-60 cm):	Brownish-yellow (10 YR 6/6) silt loam, becoming reddish-yellow (7.5 YR 6/6) with depth; blocky structure, better developed with depth, very porous; calcareous pebbles; gradual to diffuse boundary.
(B)C:	Pale brown colluvial silt loam, with more calcareous gravel than above.

Geochemical and Biochemical Properties

Particle-size distribution: Texture is finer than in the Black or Brunified Rendzinas observed at higher elevations. There is very little sand (3-4%) and 17% clay. Clay is not eluviated.

Exchange complex and carbonates: Less $CaCO_3$ [4-5% in (B), of which 50% is active] than in Humic Rendzinas. Decarbonation is almost complete in A₁; yet the exchange complex remains saturated (mainly with Ca^{2+}) and pH is close to neutral [6.8 in A₁ and 7.2 in (B)].

Biochemical properties: Organic matter content is still high in A₁ (6% in the upper part of this horizon), but lower than in Rendzinas. Organic matter decreases through the mineral horizon at a faster rate than in Brunified Rendzinas. Very active mull, with a C/N ratio of 12 in A₁.

Iron and aluminum oxides: The anomalously high content of free iron (3%) and even of aluminum (0.7%) are signs of the "polycyclic" origin of this Colluvial soil. These elements are inherited from the terra fusca. In contrast with Rendzinas, the organic matter content of this slightly calcareous soil is not high enough to mask the color induced by iron.

Genesis. In catenas found along the slopes of dry dells carved in limestone plateaus, these soils develop on the valley floor, whereas Humic and Brunified Rendzinas are observed at the foot of the slopes, on the coarsest colluvium. Soil climate is much cooler than that of Rendzinas. The abundance and the variety of the vegetation (especially ash) are indicative of a rich chemical medium and of good physical conditions. Compared to Rendzinas, the parent material is richer in fine silicate particles and lower in carbonates. The solum is deeper and contains less coarse fragments. As the surface organic matter is no longer immobilized by carbonates, mineralization prevails over humification. Consequently, the A horizon is thinner and lower in humus, while the (B) horizon is more developed and has a more pronounced blocky structure. These features can be used to differentiate Calcareous Brown soils from Brunified Rendzinas. In addition, because of the presence of carbonates in the (B) horizon, this soil contrasts with Calcic Brown soils and with Eutrophic Brown soils, although they are all polycyclic and formed under similar conditions (terra fusca mixed with calcareous gravel by cryoturbation or colluviation).

REFERENCE: Gury, M., *Legend, Soil Map of the Haye Plateau,* C.N.R.S., Nancy, 1972.

VI₆: CALCIC BROWN SOIL

(Sol brun calcique)

F.A.O.: *Calcic Cambisol;* U.S.: *Typic Eutrochrept*

Location: Agricultural Research Station, Dijon, Côte-d'Or, France.
Topography: Level land. Elevation 269 m.
Parent material: Upper Oligocene sandy marl.
Climate: Atlantic with continental trend. P. 700 mm; M.T. 10.5°C.
Vegetation: Plowed land.

Profile Description

A$_p$ (0-20 cm):	Dark yellowish-brown (10 YR 4/4), moderately humic silty clay; crumb to blocky structure; some quartz or limestone pebbles; few fine roots; diffuse boundary.
(B) (20-45 cm):	Dark yellowish-brown (10 YR 4/4), weakly humic silty clay; coarse blocky structure; some pebbles and gravel; few fine roots and worm-holes; slightly effervescent.
(B)C (45-65 cm):	Strong brown (7.5 YR 5/8) weathered marl; blocky to massive structure; strongly effervescent; clear boundary.
C$_{ca}$ (65-75 cm):	Yellow (10 YR 7/8), very calcareous marl, loam texture; white calcareous nodules.
C:	Yellow marl, sandy loam texture; very calcareous, friable.

Geochemical and Biochemical Properties

Particle-size distribution and carbonates: The silty clay texture (40% clay) in A and (B) changes to loam in C$_{ca}$ with 21% clay; this results from a relative concentration of clay in the surface horizon through decarbonation. CaCO$_3$ represents 1-1.5% at the surface, increases to 34% in (B)C and to about 60% in C. However, *there is no active carbonate in the upper horizons.*

Exchange complex: The exchange capacity is high (25 meq/100 g) in A and (B) but decreases with depth. Base saturation is 100%, mainly due to Ca^{2+}. pH (water) is 7.8-8.

Biochemical properties: Because of cultivation, the organic matter content is low (2.3%) in the A$_p$ horizon, with a C/N ratio of 9.6. The (B) horizon still contains 1.4% organic matter.

Free iron: Like clay, and for the same reason, iron is concentrated in the surface horizon (1.5% in A$_p$ versus 0.5% in C).

NOTE. The (B) horizon has a bulk density of 1.2 g/cm^3 and a porosity of 52%.

Genesis. The very small amount of carbonates in the A and (B) horizons, present as coarse particles, and the absence of active carbonate result from cultivation. Under these conditions, it is very likely that decarbonation was almost complete down to 50 cm before cultivation was started. This is, thus, a Calcic Brown soil, not a Calcareous Brown soil. On the other hand, this soil is monocyclic as it results from direct and relatively fast weathering of the parent material by decarbonation together with a surface accumulation of silicates (especially clays) and iron oxides. This monocyclic evolution is evidenced by the presence of a "ca" horizon with small white concretions, resulting from the precipitation of leached carbonates in the C horizon.

REFERENCE: *Soil Map of France (1:100,000),* Dijon sheet, Agriculture Research Station, I.N.R.A., Dijon.

PLATE VI

Brunified Calcimagnesian Soils

CHAPTER IV

ISOHUMIC SOILS AND VERTISOLS

These soils are characterized by a strong "climatic maturation" (linked to the existence of a dry season) of part of the humus substances which are stabilized, polymerized, and become dark in color. These dark humus compounds, which are very resistant to microbial biodegradation, penetrate deeply into the profile and give it the appearance of an A_1C profile (isohumic process). This process characterizes two classes of soils: "Isohumic soils and Vertisols".*

Important differences exist between Isohumic soils and Vertisols:

(1) In Isohumic soils, *sensu stricto*, the labile organic matter fraction (with fast turnover) is still relatively important. The deep penetration of the organic matter is essentially due to *biological* causes, i.e., in place decomposition of the roots of the steppe grasses and intense activity of earthworms and other burrowing animals.

NOTE: Geomorphological investigations have shown the importance, in some cases, of former water tables in the formation of Chernozems (Kovda & Dobrovolsky, 1974).

(2) In Vertisols, the labile organic matter fraction with fast turnover is small. Existing only at the surface, it often disappears and is renewed from year to year. The stabilized organic matter is fixed by swelling clays which are generally very abundant. Moreover, the deep incorporation of stabilized humus substances (dark in color) is of mechanical origin and is due to "vertic movements," associated with the alternate swelling and shrinking of clay minerals.

Both classes of soil, which are clearly separated in most world taxonomies, have unquestionably some common features linked to a dry climate or at least to a well-defined dry season. They are both characterized by a specific evolution of the organic matter, the existence of a certain amount of 2/1 clay minerals of neoformation, and a high exchange capacity. The exchange complex is saturated with divalent ions (Ca^{2+} and Mg^{2+}), whose role as a "driving force" in the development of these soils is indisputable.

Isohumic soils, *sensu stricto*, result from an essentially bioclimatic evolution. For example, in Russia and in Siberia, they characterize one or even several of the *zones* defined by Dokuchaev.

*In this book, the term "polymerization" is used in the sense of "polycondensation of aromatic nuclei."

51

On the contrary, formation of Vertisols is conditioned by a *site factor* in addition to the climatic factor. This factor leads to *poor drainage*, which favors extremely contrasted seasonal soil climatic conditions. Typical features then develop such as very important neoformation (or conservation) of 2/1 clay minerals, a particular structure associated with vertic movements, and a strong maturation of the stable humus fraction.

ISOHUMIC SOILS WITH BIOCLIMATIC EVOLUTION

These soils are particularly well represented in the southern part of the U.S.S.R., where they constitute the great Chernozem zone (subdivided in several subzones) and the Chestnut soil zone. These are essentially *steppe* soils. They are characterized by an extremely contrasted soil climate which is very cold in winter and relatively warm and dry in summer. However, other types of Isohumic soils form under other climates, either more humid (Brunizems) or warmer (Reddish Chestnut soils and Tropical Isohumic soils). I shall recapitulate here only the main definitions; the reader is referred to *"Précis de Pédologie"* for more details (Duchaufour, 1970).

Main types of Isohumic soils

Chernozems. These black and base-saturated soils are characteristic of the "dense" steppe. Some transitional forms, of which examples will be given, are found in the more humid border zones where the forest begins to have an influence (Humic or Eluviated Chernozems). They also exist in the southern steppes with milder climate (Danubian Chernozems). The profile is of A_1Ca type. In climatic soils, the humic A_1 horizon is at least partially decarbonated at the surface. "Calcareous" Chernozems, of which an example will be given, are associated with particular site characteristics.

Chestnut soils. Such soils are found in the "sparse" steppe with drier climate. They have a lower organic matter content and are often incompletely decarbonated. The exchange complex contains some sodium. (In fact, site factors may be responsible for the frequency of occurrence of Sodic soils within the Chestnut soil zone.)

Brunizems. These are prairie soils with clusters of trees, under more humid continental climate; exchange complex slightly desaturated; presence of an argillic horizon (A_1B_t profile). Note the absence of a calcic horizon.

Reddish Chestnut soils (Gerasimov, 1956). Such soils have a less clearly expressed isohumic character. They are rich in reddish free iron oxides (noncomplexed), characteristic of Mediterranean or subtropical climates with a xerophilous shrubby vegetation of the *"maquis"* or *"garrigue"* type.* These soils form transition with *Fersialitic soils* occurring

---

*Translators' note: In Mediterranean regions, *maquis* designates a dense plant association of shrubs (holm oak, cork oak), myrtle, heath, etc., while *garrigue* designates a secondary plant formation (holm oak with various shrubs and grasses) developed after destruction of the forest.

under warm and more humid climates (dense xerophilous forest). Most Reddish Chestnut soils display a well-developed calcic horizon which is sometimes petrocalcic (calcareous crust); those of North Africa are generally polycyclic.

Finally, I shall describe a *Tropical Isohumic soil*, characteristic of a tropical climate with a marked dry season.

As aridity increases, Isohumic soils gradually change into semidesert soils. Gray soils, also called *Sierozems*, cannot be classified with Isohumic soils and constitute a separate class in most soil taxonomies. One example will be described in Chapter X.

Ecological evolution of Isohumic soils

Table 4 shows the evolution of Isohumic soils as influenced by temperature and humidity. It stresses the particular importance of the humidity factor, or more precisely of "climatic drainage" (precipitation-potential evapotranspiration), in determining the type of vegetation and the genetic features of the profile.

For each climate, a *characteristic humidity threshold* exists, below which carbonates migrate and accumulate in a calcic horizon and above which clays (and iron) migrate and form an argillic horizon, while the major portion of carbonates is leached out of the profile.

Such a succession is especially well illustrated by the various zones encountered from south to north in the U.S.S.R. First, the density of the steppe increases, then woody vegetation takes precedence, initially as "clumps" of trees, then as copses, and finally as continuous forest. At the same time, calcium carbonate accumulates at increasing depths, until it disappears below the soil profile. Organic matter content increases to reach a maximum in Humic Chernozems. Finally, clay migration follows that of carbonates and increases in importance as the forest takes over. A (B) horizon appears first, followed by an argillic B_t horizon.

SOILS WITH PEDOGENESIS INFLUENCED BY SITE FACTORS: VERTISOLS
(Table 5)

These intrazonal soils (site climaxes) occur not only in the climatic zone of Isohumic soils, but also in zones with a forest climax, subject to the sole climatic condition that there be a marked dry season (for example, climatic zone of Fersialitic soils). As a rule, they are not found in the humid temperate zone, where only transitional soils can form under very specific circumstances.

Those *site factors* (most often *topography* and *parent material*) which, compared to Isohumic soils, tend to accentuate the seasonal variations of the soil climate are of great importance. Vertic soils develop on Ca- and Mg-rich material and are often located in depressions with poor internal or external drainage. Seasonal contrasts—waterlogging during the wet season and strong desiccation during the dry season—combine with a high concentration of divalent cations to promote (1) neoformation or conservation of swelling clays and (2) maturation and polymerization of part of the organic matter which is tightly bound to swelling clays.

Ecological classification of Vertic soils

It is based on two main criteria: (1) the degree of development and (2) the origin and nature of the clay minerals.

Degree of development. *Vertic soils* will be opposed to *Vertisols*, *sensu stricto*. Vertic soils are less developed than Vertisols, because they display only those vertic features that are acquired rapidly, especially *structural* properties (shrinkage cracks, slickensides). Clay minerals do not consist exclusively of true montmorillonite but may include a fairly high proportion of partly swelling interstratified minerals, or even kaolinite under warm climate. The organic matter has usually not acquired vertic characteristics.

Vertisols, *sensu stricto*, are characterized by the almost exclusive presence of true montmorillonite. They show pronounced vertic properties, involving also the organic matter (black color).

Origin of clay minerals. Most Vertisols, *sensu stricto*, are characterized by *clay minerals of neoformation* (Nguyen Kha, 1973). These are always well-developed soils with respect to the origin of the weathering products. *Topomorphic Vertisols* form on basic volcanic or crystalline rock but at specific topographic positions with poor external drainage. Vertic Eutrophic Brown soils (Profile XVII₂, Chapter IX) belong to this category although they are only slightly colored by stabilized and incorporated organic matter.

On the contrary, *lithomorphic Vertisols* are relatively young soils formed in sedimentary marl from which clay minerals are inherited. In this case, the vertic development requires less emphasized topographic and soil climate conditions. Some external drainage due to a very gentle slope may even exist. However, "stabilization" of the organic matter and blackening of the profile will only occur when topography reinforces poor drainage conditions (see Profile VII₆, this chapter).

NOTE. Some Vertisols containing a larger amount of labile organic matter at the surface may even display an upper horizon with a considerably improved "crumb" structure (Grumosols). If this organic horizon is destroyed or eroded, coarse prismatic structure begins right at the surface.

Evolution of Vertisols

Vertic soils will be opposed to *Vertisols* because the former occur under a much wider range of conditions than the latter (Table 5).

From an ecological point of view, it is possible to establish subdivisions based on general bioclimatic factors (intensity of the dry season) on the one hand, and on site factors (*soil climate*—topography and parent material) on the other.

Climates with marked dry season. Included here are tropical or dry subtropical climate zones, as well as the driest climatic zones of Isohumic soils under steppe or xerophilous shrubby vegetation (see Profile III$_3$). The best examples of Vertisols are precisely found in the tropical savanna zone and in the Mediterranean Fersialitic soil zone, when local site factors permit their development (Paquet, 1969; Bocquier, 1973).

Either true Vertisols or Vertic soils forming transition towards soils with climatic genesis (Vertic Fersialitic soils and Vertic Calcareous Brown soils, Profile VI$_4$, Chapter III) will be found depending on the suitability of site factors.

Transitional climate with moderate dry season. Humid tropical climate with forest climax. The Vertisol stage with black color is rare. This is the zone of Tropical Vertic Eutrophic Brown soils (Profile XVII$_2$, Chapter IX).

Humid temperate climates with a short or no dry season. As a rule, vertisolization does not occur in these climatic zones. However, Vertic soils are found instead of recently developed Vertisols. Their characteristics are inherited either from the parent material (Vertic Pelosols developed on marl with partly swelling interstratified clay minerals, Profile XV$_1$, Chapter VII) or from a former evolution under more contrasted climate (polycyclic soils such as certain Vertic Alluvial soils).

NOTE: "Vertic soils" have not been included in this chapter but in the various chapters describing those soils with bioclimatic genesis with which they form transition.

Table 4.   Isohumic Soils: Bioclimatic Genesis

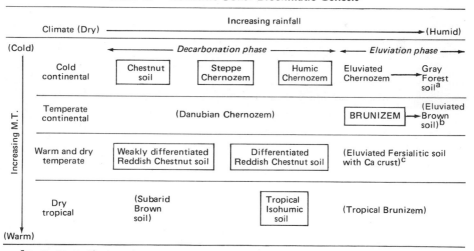

[a] See Profile X$_6$ (Chapter V).
[b] See Profile IX$_5$ (Chapter V).
[c] See Profiles XVI$_5$ and XVI$_6$ (Chapter VIII).

Table 5.   Vertic Evolution Related to Site Factors

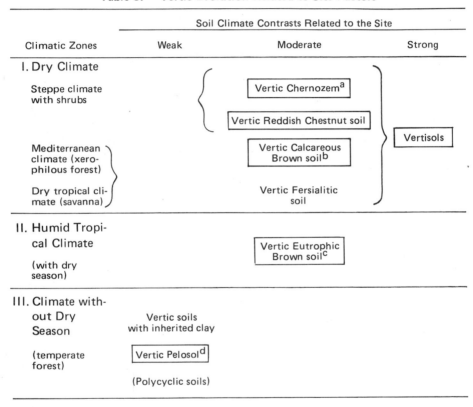

Climatic Zones	Soil Climate Contrasts Related to the Site		
	Weak	Moderate	Strong
I. Dry Climate			
Steppe climate with shrubs		Vertic Chernozem[a]	
		Vertic Reddish Chestnut soil	Vertisols
Mediterranean climate (xerophilous forest)		Vertic Calcareous Brown soil[b]	
Dry tropical climate (savanna)		Vertic Fersialitic soil	
II. Humid Tropical Climate (with dry season)		Vertic Eutrophic Brown soil[c]	
III. Climate without Dry Season (temperate forest)	Vertic soils with inherited clay Vertic Pelosol[d] (Polycyclic soils)		

[a]See Profile III$_3$ ("Vertic Alluvial soil") (Chapter II).
[b]See Profile VI$_4$ (Chapter III).
[c]See Profile XVII$_2$ (Chapter IX).
[d]See Profile XV$_1$ (Chapter VII).

VII₁: BRUNIZEM

(Brunizem)

F.A.O.: *Luvic Phoeozem*; U.S.: *Typic Argiudoll*

Location: Saybrook, northwest of Urbana, Ill., U.S.
Topography: Top of very gentle slope (1-2%).
Parent material: Loess (70 cm thick), overlying calcareous moraine deposit.
Climate: *Continental* temperate, moderately humid. P. 870 mm; M.T. 11°C.
Vegetation: "Prairie" climax, less xerophilous than the steppe; grasses and clumps of trees; cornbelt.

Profile Description

A_1 (0-43 cm):   Black (10 YR 2/1) humic silty clay loam, becoming very dark gray (10 YR 3/1) with depth; crumb structure, friable, well aerated; common grass roots; gradual boundary.

B_t (43-71 cm):   Dark grayish brown (10 YR 4/2) silty clay loam becoming brown (10 YR 5/3); gradually coarser blocky structure; dark gray (10 YR 4/1) clay skins; firm; abrupt boundary.

(B)C (71-90 cm):   Yellowish-brown (10 YR 5/4) to light yellowish-brown (10 YR 6/4) loam; coarse blocky structure; more evident dark gray (10 YR 4/1) clay skins.

C (90-107 cm):   Dark grayish-brown (2.5 Y 4/2); very firm; resting on effervescent calcareous material.

Geochemical and Biochemical Properties

Particle-size distribution: "Homogeneous" silty clay loam to the 70-cm depth (5% sand, more than 50% silt). Moderate clay eluviation (29% clay in A_1 and 39% in B_t). Texture becomes lighter in (B)C and C.

Clay minerals: 2/1 type minerals consisting mainly of montmorillonite and illite, some chlorite. Montmorillonite is clearly dominant in the B_t horizon.

Exchange complex: Relatively high exchange capacity (27 meq/100 g in A_1 and 28 meq/100 g in B_t). The exchange complex is slightly desaturated (89% in A_1 and 85% in B_t, pH in water 5.7). Exchangeable Mg^{2+} is very abundant (9 meq/100 g) with a Ca^{2+}/Mg^{2+} ratio of 1.6

Biochemistry: Organic matter content in A_1 is 5% at the surface and reduces to 4% at its base. Presence of organic matter in the deep horizons [2% in B_t and (B)C] which gives the clay skins a gray color.

Genesis. Like Chernozems, Brunizems are "Mollisols" that are characterized by the predominance of 2/1 clay minerals, an abundance of Ca^{2+} and Mg^{2+} on the exchange complex, and a strong humification of the humic A_1 horizon. The humic horizon is, however, thinner and slightly more desaturated than in Chernozems. It is also strongly acid. Finally, humic acids are less strongly polymerized and have undergone a less intense "climatic maturation." This reflects the influence of a milder and more humid climate and opposes the "prairie" to the "steppe." The woody element of this humid continental plant formation is nonnegligible. Scattered copses are likely to change place if fire occurs. The formation of a very special type of "argillic" horizon with weak vertic properties can be attributed to the action of trees. The dark gray coatings in B_t are the result of a mechanical translocation of fine clay minerals (montmorillonite) and of organic matter from the humic horizons. To the north, this type of soil gives way to Eluviated Brown soils and Eluviated soils which are less humic at the surface and are essentially of forest origin.

REFERENCE: *7th International Congress of Soil Science*, Madison, Wisconsin, U.S., 1960.

VII₂: HUMIC CHERNOZEM (typical)

(Chernozem humifère)

F.A.O.: *Calcic Chernozem;* U.S.: *Pachic Haploboroll*

Location: Karlinsky State Farm, Oulyanovsk, U.S.S.R.
Topography: Level plain. Elevation 130 m.
Parent material: Reworked periglacial loam.
Climate: Cold and dry continental. P. about 400 mm; M.T. 3.5°C (Jan. −13°C, July +20.5°C).
Vegetation: Cultivated land. Former native vegetation was steppe in the forest-steppe zone, with 40% forest cover.

Profile Description

A_p (0-25 cm):   Very dark gray (10 YR 3/1) humic loam; compact peds, hard when dry; few roots; clear smooth boundary.

A_1 (25-45 cm):   Very dark gray (10 YR 3/1) humic loam; granular structure, well aerated; porous; many fine roots; gradual boundary.

A_1 (B) (45-70 cm):   Dark grayish-brown (10 YR 4/2) clay loam; subangular blocky structure; krotovinas (burrows filled with humus); gradual boundary.

(B)C (70-110 cm):   Yellowish-brown (10 YR 5/6) clay loam; strong prismatic to blocky structure; no cutans, humic krotovinas.

C_{ca} (110-240 cm):   Yellowish-brown clay loam; massive structure; diffuse and weakly expressed pseudomycelium.

Geochemical and Biochemical Properties

Particle-size distribution: The parent material is a clay loam with 27-28% clay and 4% $CaCO_3$. Slight and gradual translocation of clay can be observed (22% clay in A_p). The illuviated clay is so dispersed that "clay skins" do not form. Clay minerals consist of montmorillonite and illite in approximately equal amounts.

Exchange complex: The high exchange capacity (50 meq/100 g) is characteristic of both the nature of the clay and the humus. Ca^{2+} predominates while Mg^{2+} represents 2-3 meq/100 g. Na^+ and K^+ are present as traces. Carbonates can be detected in small amounts from 50 to 60 cm and $CaCO_3$ content reaches 15% in the C_{ca} horizon with pseudomycelium.

Biochemistry: The A_p and A_1 horizons have almost the same organic matter content (7%) and a C/N ratio of 13.5. Humic acids are more abundant than fulvic acids (FA/HA = 0.3) and consist mainly of gray humic acids with very polymerized nuclei. Nonextractable humin (even after elimination of iron) still represents 35 to 46% of total carbon.

Iron and aluminum hydroxides: The high concentrations of free iron (3%) and aluminum (1%) extracted with the Tamm reagent are due to insoluble complexed forms.

Genesis. This Chernozem is typical of the northern forest-steppe zone with colder climate, where the steppe (on plateaus) alternates with forest covering the slopes and hills. With a very dark A_1 horizon, often thicker than 1 m, this type of Chernozem has the highest organic matter content of all Chernozems. This soil is close to an Eluviated Chernozem. The (B) horizon has a strongly expressed structure, but it cannot yet be qualified as an "argillic" horizon as signs of eluviation are not very apparent. On the other hand, it contrasts with the Steppe Chernozem (Profile VII₃) because of its greater organic matter content, its darker color, the presence of a (B) horizon, and the generally more intense decarbonation at the surface. However, accumulation of carbonates in C_{ca} is less discernable. The intense polymerization of the humus due to prolonged maturation under cold climate with contrasting seasons is a characteristic of all Chernozems. The average age obtained by ^{14}C dating is several thousand years (Gerasimov, 1974b).

REFERENCE: Ivanov, *Guidebook to field trip 8, 10th International Congress of Soil Science,* Moscow, 1974. (Photo by F. Jacquin.)

VII₃: STEPPE CHERNOZEM (orthic)

(Chernozem de steppe)

F.A.O.: *Calcic Chernozem*; U.S.: *Typic Calciboroll*

Location: 50 km south of Zaporojie, U.S.S.R.
Topography: Plateau dissected by asymmetrical valleys. Elevation about 130 m.
Parent material: Loess containing carbonate, 20 m thick.
Climate: Cold continental. P. 450 mm; M.T. Jan. $-7°C$, July $+21.5°C$.
Vegetation: Dense steppe with *Stipa lessingiana*, *Festuca sulcata*, *Koeleria gracilis*, etc. Currently under cultivation.

Profile Description

A_1 (0-40 cm): Very dark gray (10 YR 3/1) silty clay loam; irregular medium crumb structure, disturbed at the surface by cultivation (A_p); well aerated; many roots; gradual boundary.

A_1/C (40-60 cm): Dark grayish-brown (10 YR 4/2) silty clay loam; fine granular structure; effervescent below 55 cm; presence of krotovinas (burrows filled with humus from A_1); many roots; gradual boundary.

C_{ca} (60-100 cm): Grayish-brown (10 YR 5/2) silty clay loam, becoming gradually lighter; coarser structure; few roots; accumulation of $CaCO_3$ as powdery spots and friable concretions.

C: Yellowish-brown compact loess; massive structure.

Geochemical and Biochemical Properties

Particle-size distribution: The silty clay loam texture is uniform throughout the profile (36-38% clay and very little sand). Clay minerals are mainly "inherited" from the loess and dominated by illites. Montmorillonite is low in A_1 but increases in C_{ca}.

Exchange complex: The high exchange capacity of 38 meq/100 g is lower than in the Humic Chernozem. The profile is base-saturated, mainly with calcium. Mg^{2+} contributes 5-8 meq/100 g. pH is 7.5 at the surface. $CaCO_3$ is present at 55 cm and represents 17% in C_{ca} (spots and concretions; pH 8).

Biochemistry: The organic matter content of Chernozems decreases from north to south. It is therefore lower in this profile (5% in A_1, decreasing to 4%, and 2.5% in A_1/C) than in the Humic Chernozem. The C/N ratio of 11-11.5 is quite uniform throughout the profile, but lower than in Profile VII₂ which is partly subjected to the forest influence. This soil has less than 50% extractable carbon, a high content of very mature humin and a low FA/HA ratio of about 0.40. These are usual characteristics of the highly mature humus of the steppe.

Hydroxides: Free iron hydroxide content is not known. Iron content in the clay fraction is exceptionally high at 13.5% and indicates the presence of ferriferous clay minerals.

Genesis. This type of Chernozem characterizes the still dense steppe of the southern part of the Ukraine. It succeeds the Humic Chernozem and precedes the southern Chernozem. The organic matter content decreases, the (B) horizon disappears, and the accumulation of $CaCO_3$ becomes more evident as spots and concretions replace the diffuse pseudomycelium. Compared to northern plant formations where trees play a nonnegligible role, the steppe accumulates less humus, mobilizes less calcium carbonate and does not induce the formation of a (B) or a structural B horizon since roots remain near the soil surface. However, the bioclimatic maturation of humus is fundamentally similar throughout the whole Chernozem zone with cold continental climate.

REFERENCE: *Field trip to the Ukraine-Crimea, 10th International Congress of Soil Science,* Moscow, 1974. (Photo by F. Le Tacon.)

VII₄: CALCAREOUS CHERNOZEM (subcaucasian)
(Chernozem calcaire)
F.A.O.: *Haplic Chernozem*; U.S.: *Calcic Haploxeroll*

Location: Gigant Sovkhoz, 140 km east of Rostov-on-Don, U.S.S.R.
Topography: Level. Elevation about 170 m.
Parent material: Windblown clay loam.
Climate: Dry continental. P. 400-450 mm; M.T. 8.7°C (Jan. −5°C, July +25°C).
Vegetation: Native vegetation is dry steppe. Cereal crops protected by tree windbreaks.

Profile Description

A_p (0-25 cm): · Very dark grayish-brown (10 YR 3/2) silty clay loam; blocky structure disturbed by cultivation; slightly effervescent; clear boundary.

A_1 (25-40 cm): Very dark grayish-brown (10 YR 3/2) silty clay loam; strong crumb structure; effervescent; common roots; gradual boundary.

A_{1ca} (40-75 cm): Very dark grayish-brown becoming lighter with depth; compact subangular blocky structure; diffuse and weakly expressed pseudomycelium; gradual boundary.

(B)C (75-130 cm): Yellowish-brown (10 YR 5/4) silty clay loam with vertical humic tongues; compact cubic to prismatic structure; gradual boundary.

C_{ca} (130-200) cm): Light yellowish-brown (10 YR 6/4) silty clay loam; cubic to prismatic structure; well-defined accumulations of white and friable $CaCO_3$.

Geochemical and Biochemical Properties

Particle-size distribution: Silty clay loam with less than 2% sand. Uniform clay content (37-38%) comprised mainly of montmorillonitic clay minerals.

Exchange complex: The high exchange capacity (35 meq/100 g) is characteristic of the type of clay minerals and organic matter found in the profile. The exchange complex is base-saturated (pH 7.1 at the surface increasing to 8.2 in C_{ca}). Dominant exchangeable cations are Ca^{2+}, then Mg^{2+}, but Na^+ may amount to 10% in the A_{1ca} horizon of some profiles. This marks a transition towards Solonetz. $CaCO_3$ *is present in all horizons* and increases from 1.5% at the surface to 8% in C_{ca}.

Organic matter: Organic matter content is 4% in the A_p and A_1 horizons, still 3.5% in A_{1ca}, and 2% in (B)C. This is very dark organic matter whose humic components (gray humic acids and humin) are characterized by polymerized nuclei. The C/N ratio ranges from 11 to 10.

Iron and aluminum hydroxides: Total iron content is high (6%). Thus, most of the iron is included in the clay lattice and in the organic matter as insoluble and stable complexes.

Genesis. Calcareous Chernozems are characteristic of the low plains of the southern U.S.S.R. with a dry and relatively warmer continental climate. The native vegetation consists of a much less xerophilous and denser steppe than the vegetation of Chestnut soils of the Volga-Don region. On the other hand, the warmer and less contrasted climate than that of the Ukraine results in a lower accumulation of humus (see Danubian Chernozem, Profile VII₅). Some peculiarities of the profile, especially the presence of $CaCO_3$ in all horizons, can be explained by two factors: (i) the low permeability of the parent material, which contains more clay than usual, slows decarbonation at the surface; and (ii) the geomorphological evolution whereby, according to Kovda, a water table was present during some periods of the Quaternary and favored the capillary "rise" of carbonates. Compared to modal Steppe Chernozems, the deep incorporation of organic matter is related to these same two factors.

REFERENCE: *Guidebook to the Volga-Don field trip, 10th International Congress of Soil Science,* Moscow, 1974.

VII₅: ELUVIATED DANUBIAN CHERNOZEM

(Chernozem danubien lessivé)

F.A.O.: *Luvic Chernozem;* **U.S.:** *Vertic Argiustoll*

Location: Unirea region, Transylvania, Romania.
Topography: Level. Elevation 310 m.
Parent material: Silty clay resting on fluvial terrace.
Climate: Temperate continental. P. 540 mm; M.T. 9.5°C.
Vegetation: Forest-steppe zone with predominating forest (common oak). Under cultivation at this site.

Profile Description

A_1 (0-40 cm): Very dark gray (10 YR 3/1) clay loam; well aerated crumb structure, more massive at the surface due to cultivation (A_p); friable when moist, hard when dry; common roots, gradual boundary.

A_1/B_t (40-75 cm): Very dark gray (10 YR 3/1) clay, becoming lighter with depth; blocky to prismatic structure; shrinkage cracks when dry; clay skins on peds; gradual boundary.

B_t (75-110 cm): Brown (10 YR 5/3) clay; strong blocky to cubic structure; deep penetration of vertical humic tongues; shiny coatings; gradual boundary.

C_{ca}: Brown silty clay; massive structure; presence of carbonates below 170 cm as pseudomycelium and calcareous spots.

Geochemical and Biochemical Properties

Particle-size distribution: Silty clay with 41-43% clay in C. Evidence of clay translocation from A_1 to A_1/B_t and B_t. Clay content is 40% in A_1, 59% in A_1/B_t, and 52% in B_t. Like all soils with a vertic tendency, clay accumulation is very dispersed at depth and limited to the faces of shrinkage cracks. Clay minerals consist of illite and montmorillonite. The latter could result from moderate neoformation in A_1 and A_1/B_t.

Exchange complex: The kinds and the quantity of clay minerals are responsible for a high exchange capacity (38 meq/100 g in A_1A_p and 48 meq/100 g in A_1/B_t). In the C_{ca} horizon, a lower content of montmorillonite results in a lower exchange capacity of 25 meq/100 g. The profile is slightly desaturated at the surface (S/T 84%; pH 6.6). But S/T increases to 100% in C_{ca} which contains 24% $CaCO_3$ below 230 cm. The dominant cations are Ca^{2+} (68%) and Mg^{2+}, with only traces of Na^+.

Biochemical properties: Organic matter content is 4% in A_1 and still 2.5% in A_1/B_t. A C/N ratio of 11 indicates intense biological activity.

Genesis. Eluviated Chernozems characterize the "forest-steppe" zone dominated by the forest, at the edge of the zone dominated by the "steppe." There, the various types of noneluviated Chernozems are found. The influence of trees is manifested by the eluviation and the dispersed accumulation of clay in a deep and well-structured B_t horizon and through the slight desaturation of the exchange complex at the surface. The "isohumic" character of the profile is in turn caused by the influence of grasses. This soil forms transition with the "Gray Forest" soils of the U.S.S.R. in the north, but with Danubian Chernozems in the south, as is the case here. Therefore, the organic matter of this soil displays similarities with that of the noneluviated Danubian Chernozems in opposition to the Chernozems with cold continental climate found in the Ukraine. The smaller seasonal variations in temperature produce a less complete maturation and stabilization of the humus, although seasonal biodegradation remains important. The stable and permanent fraction of the humus is less abundant and lighter in color than in the Ukraine Chernozems.

REFERENCE: Popovatz, M., A. Popovatz, and C. Rapaport, *Guidebook to field trip 2. 8th International Congress of Soil Science*, Bucharest, 1964.

VII₆: VERTISOL

(Vertisol)

F.A.O.: *Pellic Vertisol;* U.S.: *Typic Pelloxerert*

Location: Guadalquivir Valley, along the Cordova to Sevilla road, Spain.
Topography: Depression surrounded by hills. Elevation about 80 m.
Parent material: Miocene (Burdigalian) marl.
Climate: Semi-arid Mediterranean. P. 500 mm; M.T. 19.6°C (at Sevilla).
Vegetation: Cotton crop.

Profile Description

A_p (0-25 cm): Very dark gray (10 YR 3/1) clay; friable crumb structure; few roots; moderate biological activity; gradual boundary.

A/B (25-50 cm): Transitional horizon with gradually coarser structure.

$(B)_1$ (50-100 cm): Very dark gray (10 YR 3/1) clay; coarse prismatic structure, wide cracks; firm and plastic; few calcareous concretions, weakly expressed slickensides.

$(B)_2$ (100-150 cm): Same characteristics as above, except blocky structure with numerous diagonal slickensides.

C: Grayish-brown marl.

Geochemical and Biochemical Properties

Particle-size distribution: Clay content is high and very uniform throughout the profile with 54% swelling clay minerals, comprised mostly of montmorillonite and some illite. Quartz sand (30%) is inherited from the parent material. $CaCO_3$ content is 15%, as in the parent material.

Exchange complex: High exchange capacity (60 meq/100 g of soil or over 100 meq/100 g of clay). The exchange complex is base-saturated with Ca^{2+} (about 55 meq/100 g) and Mg^{2+} (about 5 meq/100 g).

Biochemistry: Low organic matter content, uniformly distributed with 1.55% in A_p (C/N ratio 12), then about 1% over more than 1 m (C/N ratio 10). The organic matter of Vertisols is dominated by nonextractable humin and highly polymerized gray humic acids which are strongly bound to clay minerals (Nguyen Kha, 1973).

Iron: Although total iron content is not negligible (about 3%), free iron content is very low (0.10%). Thus, the weathering index is also very low at about 0.03. Montmorillonite is ferriferous. Additionally, part of the iron which is complexed by gray humic acids is not extractable.

Genesis. With respect to the mineral material, this Vertisol is weakly developed since most of the mineral constituents are inherited from the parent material without modification. Therefore, it differs from Vertisols developed on basic crystalline rock in which swelling clays result from neoformation in a calcic environment. However, the very contrasting soil climate prevailing at this site has influenced the evolution of the organic matter. Seasonal alternate wetting and drying cycles account for the "polymerization" and the "stabilization" of a small portion of the humus (black color) while the remainder is completely mineralized. Moreover, the very deep and homogeneous incorporation of the organic matter has been caused by vertic motion. Soils found on the hills and slopes surrounding the depression have not matured. These are Regosols characterized by the light color of marl.

REFERENCE: Albareda, J.M., *Conference on Mediterranean soils.* Institute of Edaphology, Madrid, 1966.

PLATE VII

BRUNIZEMS, CHERNOZEMS, AND VERTISOLS

VIII₁: TROPICAL ISOHUMIC SOIL

(Sol isohumique tropical)

F.A.O.: *Calcic Kastanozem;* U.S.: *Haplic Calcixeroll*

Location: 4 km west of Garzon, along the Garzon to Neiva road, Huila, Colombia.

Topography: Platform at the foot of Upper Tertiary hills. Elevation 800 m.

Parent material: Colluviated volcanic material, consolidated into "tufa."

Climate: Semi-arid tropical. P. 900 mm with pronounced dry season; M.T. 24.4°C.

Vegetation: Savanna with Cactaceae, thorny shrubs (legumes), and xerophilous grasses.

Profile Description

A₁₁ (0-35 cm):   Very dark grayish-brown (10 YR 3/2) loamy sand; medium blocky structure, firm; many roots; clear boundary.

A₁₂ (35-60 cm):   Dark grayish-brown (2.5 Y 4/2) loamy sand, becoming lighter with depth; medium blocky structure; firm, slightly hard; few roots; strongly effervescent; diffuse boundary.

C_ca (60-100 cm):   Grayish-brown (10 YR 5/2) sandy loam; prismatic structure, very firm; white powdery pseudomycelium forming coatings.

Geochemical and Biochemical Properties

Particle-size distribution and clay minerals: Sand containing 20-30% fine particles. Clay content is lower in A₁₁ (10%) than in C_ca (16%). Clay minerals are a mixture of montmorillonite and illite (O.R.S.T.O.M. analysis).

Exchange complex and carbonates: Very high exchange capacity (15-16 meq/100 g of soil or more than 100 meq/100 g of clay). The A₁₁ horizon is decarbonated (pH 6.8; S/T 95%) but the A₁₂ and C_ca horizons are enriched with illuvial $CaCO_3$, as shown by the presence of a pseudomycelium in C_ca. In A₁₂, pH is very high and may exceed 9. Exchangeable bases are characterized by an abundance of Mg^{2+} (5.4 meq/100 g) and Na^+ (2 meq/100 g) in A₁₂ with respect to Ca^{2+} (9.5 meq/100 g).

Biochemistry: Organic matter is low (1% in A₁₁) but it consists of humin and gray humic acids with highly polymerized nuclei. This accounts for the dark color of the upper horizons.

Genesis. "Isohumic" soil, characteristic of the dry tropical savanna with Cactaceae and thorny shrubs (Mimosaceae). This soil bears some resemblance to Boreal Chernozems because of the presence of silica-rich clay minerals, a weakly differentiated A₁C profile, strongly polymerized humic compounds, and the formation of a pseudomycelium caused by the "biological" solubilization of calcium carbonate. However, the very rapid biodegradation of the labile organic matter results in a lower FA/HA ratio. In this respect, this soil is similar to the "Subarid Brown soils" of the African Sahel (Maignien, 1959). Nevertheless, the more pronounced seasonal extremes in humidity promote maturation of the humus which gives the profile a darker color. The abundance of exchangeable Mg^{2+} and Na^+ is characteristic of these soils. These cations are released during the weathering of the volcanic parent material and, due to the dry season, are incompletely leached. During the wet season, they become more or less soluble (as carbonates), thereby strongly increasing the pH. This process was described by Ruellan (1970) in Reddish Chestnut soils for which this profile shows some affinity.

REFERENCE: Faivre, P. and H. Ruiz, Codazzi Institute, Bogota, Colombia, 1974.

VIII₂: STEPPE CHESTNUT SOIL

(Sol châtain de steppe)

F.A.O.: *Calcic Kastanozem;* U.S.: *Typic Calcixeroll*

Location: Volga-Don Sovkhoz, 40 km east of Iliovka, U.S.S.R.
Topography: Undulating plain dominated by Chestnut soils (but with Solonetz in the depressions). Elevation 150 m.
Parent material: Windblown loam, 6 m thick.
Climate: Arid continental. P. 320 mm; M.T. winter $-11°C$, July $+24°C$.
Vegetation: Native vegetation is a steppe with *Artemisia, Festuca, Stipa* sp.; presently under irrigated cultivation.

Profile Description

A_1A_p (0-30 cm): Dark grayish-brown (10 YR 4/2) silt loam; quite compact with coarse peds and fine pores; clear boundary.

A_{1ca} (30-50 cm): Brown (10 YR 5/3) silt loam; compact blocky structure; effervescent; transitional horizon with gradual boundary.

C_{ca} (50-80 cm): Light yellowish-brown (10 YR 6/4) silt loam; massive structure, compact; powdery $CaCO_3$ spots, vertical humic tongues.

C (80-180 cm): Light yellowish-brown silt loam; occurrence of crystalline gypsum concretions at 150 cm.

Geochemical and Biochemical Properties

Particle-size distribution: Silt loam with 12% fine sand and 23-25% uniformly distributed clay. The fine clay minerals, often Na-saturated, have been washed down into depressions where they form an important element of Solonetz.

Exchange complex and salinity: The profile is practically devoid of soluble salts within the first 150 cm; below this depth, salinity increases gradually to about 200 cm where it becomes appreciable (formation of gypsum crystals). Calcium is the dominant cation in the soil solution. The exchange complex, saturated throughout the profile, is dominated by Ca^{2+} and Mg^{2+}, while Na^+ represents less than 7% of S. Under these conditions, this soil can be considered neither saline nor sodic. The profile is decarbonated to a depth of 30 cm. $CaCO_3$ content is 15% in A_{1ca} and increases considerably in C_{ca}.

Biochemistry: The color of the profile is due to 2-2.4% organic matter present in A_1, but polymerization is not as strong as in Chernozems (FA/HA ratio of 0.6). In A_{1ca}, the organic matter content is still 2%. Note the abundance of humin, about 80% of total organic matter, which is a sign of strong maturation of the humus.

Genesis. Chestnut soils characterize the driest steppes of European Russia, in the Volga-Don region. The humidity of the climate increases towards the Caucasus and the Black Sea where again Chernozems are found (Profile VII₄, Calcareous Chernozem). Organic matter is less abundant and less polymerized than in Chernozems due to the low density of the native vegetation and to less contrasted seasonal changes in soil humidity. Also, decarbonation takes place over a shallower depth than in Chernozems (excepting Calcareous Chernozems where local conditions play a role). Chestnut soils are often slightly saline at depth as the Na^+ ion plays a nonnegligible role. Na-saturated clay minerals concentrate mostly in depressions and are at the origin of Solonetz which are always associated with Chestnut soils.

REFERENCE: *Guidebook to the Volga-Don field trip, 10th International Congress of Soil Science,* Moscow, 1974.

VIII₃: WEAKLY DIFFERENTIATED REDDISH CHESTNUT SOIL
(with nodular crust)

(Sol marron, peu différencié)

F.A.O.: *Calcic Kastanozem;* U.S.: *Typic Calcixeroll*

Location: Saïs of Fès, Morocco.
Topography: Side of valley with gentle slope (5%); well drained. Elevation 470 m.
Parent material: Reworked clay loam.
Climate: Semi-arid Mediterranean. P. 545 mm; M.T. about 20°C.
Vegetation: *Oleo lenticetum* with *Ziziphus lotus* degraded by grazing. "Secondary" steppe.

Profile Description

A₁ (0-30 cm):   Dark reddish-brown (5 YR 3/4) calcareous clay; crumb to blocky structure; gradual boundary.

(B)$_{ca}$ (30-50 cm):   Reddish-yellow (5 YR 6/6) very calcareous clay to clay loam; coarser blocky structure than above; many calcareous granules, powdery calcareous tongues in cracks.

C$_{ca}$:   Red (2.5 YR 4/6), very calcareous; blocky to prismatic structure; calcareous nodules, increasing in size and more friable towards the base.

Geochemical and Biochemical Properties

Particle-size distribution and carbonates: Clayey material; clay content decreases with depth [49-51% in A₁ and 32% at the base of (B)$_{ca}$]. Calcium carbonate follows an inverse pattern [17% in A₁, 56% in (B)$_{ca}$, and 64% in C$_{ca}$]. Illite and montmorillonite clay minerals are dominant; probable neoformation of montmorillonite in A₁ (Boulaine, 1957).

Exchange complex: Exchange capacity not available, but the exchange complex is saturated with divalent cations (Ca^{2+} and Mg^{2+}). pH is 7.7 at the surface and more than 8 in C$_{ca}$. The abundance of Mg^{2+} often causes alkalinization at depth (Ruellan, 1970).

Biochemistry: Organic matter content decreases slightly within the A horizon (2.5% near the surface and 1.8% at the base). C/N ratio is low at 9. The polymerization of humus components is quite high, although lower than in Vertic Reddish Chestnut soils (Profile VIII₄).

Genesis. These soils have been described by Boulaine (1957) and Ruellan (1970) under the name of Reddish Brown soil with "moderate differentiation" and a "gradual accumulation" of carbonates. Gerasimov (1956) called them "Reddish Chestnut soils" and compared them to the soils found in Georgia, in the southern U.S.S.R. Similarities between the genesis of these two geographically distinct groups seem certain, mainly from an ecological point of view. It is likely, however, that the Georgian profiles have undergone a less complex evolution. In the Reddish Chestnut soils of North Africa, a moderate isohumic process has been imposed on an older rubefied material. Frequently, this rubefied material has formed at high elevation under the influence of a more humid climate and under forest cover (*terra rossa*). It has then been transported to the piedmont zone with drier climate where the xerophilous shrubby vegetation, or secondary "steppe," has caused the organic matter to be deeply incorporated. The "calcic" horizon, enriched with carbonates which originate both from vertical leaching and lateral input, is still discontinuous and incompletely consolidated—powdery spots grade into harder nodules.

REFERENCE: Faraj, H. *Soil Map of Morocco*, Cah. Rech. Agron., *24*, 1967. (Photo and profile description by J. Herbauts.)

VIII₄: VERTIC REDDISH CHESTNUT SOIL (well differentiated, with calcareous crust)

(Sol marron vertique)

F.A.O.: *Calcic Kastanozem;* U.S.: *Vertic Calcixeroll*

Location: Saïs of Meknès, Morocco.
Topography: Shallow depression in a sublevel zone. Elevation 530 m.
Parent material: Reworked clay loam with pebbles over lacustrine calcareous deposits.
Climate: Semi-arid Mediterranean. P. 547 mm; M.T. about 19.3°C.
Vegetation: *Oleo lenticetum* climax. Cereal crops.

Profile Description

A₁ (0-45 cm):	Dark reddish-brown (5 YR 3/2) clay loam, noncalcareous; subangular blocky structure with prismatic trend; abrupt boundary.
(B)ca (45-75 cm):	Yellowish-red (5 YR 5/6) calcareous clay loam; blocky to prismatic structure; hard calcareous nodules and friable nodules becoming more abundant with depth.
C₁ca (75-130 cm):	Darker clay loam, very calcareous; presence of indurated granules coated with powdery accumulations becoming continuous at depth; abrupt boundary.
C₂ca:	Continuous crusting with nodules, slightly friable, compact at places.

Geochemical and Biochemical Properties

Particle-size distribution and carbonates: Clay content is relatively high: about 40% in A₁, 30% in (B)ca. Decarbonation is complete in A₁. CaCO₃ content increases *very rapidly* from 12% in (B)ca, to 30% in C₁ca, and to 70% in C₂ca.

Exchange complex: Saturated throughout even in A₁, where pH (water) is 7.8. pH reaches 8.4 in the calcareous horizons.

Organic matter: 2% in A₁ with a C/N ratio of about 10. The dark color is due to strong polymerization of the aromatic "nuclei" of the humus substances (gray humic acids). This is a vertic feature. Under native vegetation, the organic matter content would definitely be higher than under cultivation.

Genesis. Boulaine (1957) and Ruellan (1970) called this soil a well-differentiated "Red Chestnut" soil with abrupt accumulation of carbonates. Moreover, it is slightly "tirsified"—meaning it has acquired some vertic features associated with topographic position and poor drainage. The organic matter has undergone a particular evolution caused by strongly contrasting soil climatic conditions (alternate wetting and drying cycles) (see Profile VII₆, Vertisol). The complete decarbonation at the surface can also be explained by the soil climate which is locally more humid than in the preceding profile. Apart from these slight differences, the genesis of these two profiles is fundamentally the same. The successive phases of formation of a calcareous crust can be observed here by comparing the various horizons: (i) local formation of powdery spots, (ii) hardening into "nodules," and (iii) cementation of the nodules within a powdery matrix becoming gradually more compact. Carbonates are mainly supplied by lateral input (Ruellan, 1970).

REFERENCE: Faraj, H., and G. Missante, *Soil Map of Morocco*, Cah. Rech. Agron., *24*, 1967. (Photo and profile description by J. Herbauts.)

PETROCALCIC HORIZONS

The Reddish Chestnut soils of the driest climatic zones are often characterized by hardened calcareous accumulations in some horizons which are commonly called "calcareous crusts" or more scientifically "petrocalcic horizons." Two examples are given in Profiles VIII$_5$ and VIII$_6$.

The soils of Plate VIII illustrate the manner by which carbonates accumulate in soils under steppe or shrubby vegetation under dry climate. Calcic horizons can be classified according to: (i) the amount of carbonates, which form, at first, dispersed or localized accumulations, and, at a later stage, a continuous mass; and (ii) the degree of cohesion of accumulated carbonates which are soft (powdery) at the beginning or very hard and crystalline at later stages of development. Both criteria are related to the age of the accumulation, on the one hand, and to the contrasts in soil climate of the "ca" horizon, on the other. Hardening of the crust increases as seasonal changes in soil humidity become more contrasted.

The following can be distinguished:

1. A diffuse pseudomycelium (Profile VIII$_1$)
2. Powdery spots (Profile VIII$_2$)
3. Powdery spots in the process of hardening (nodules) (Profile VIII$_3$)
4. Cemented nodules forming a crust (Profile VIII$_4$)
5. A massive and platy crust (Profile VIII$_5$)
6. A layered hardened crust (Profile VIII$_6$)

In all cases, lateral leaching of carbonates plays a considerable role, especially in the formation of thick crusts. Profiles with a calcareous crust are most often located in piedmont zones, at the bottom of slopes (Ruellan, 1970).

VIII₅: REDDISH CHESTNUT SOIL WITH PLATY CRUST

(Sol marron à encroûtement feuilleté)

Jebel Lahmar, Morocco

This crust is massive, very thick, soft, and tufaceous. Bands of harder material are interleaved diagonally.

Genesis. It may be assumed that this crust, located at the foot of a slope in the piedmont zone, is of colluvial origin. It could result from the cementation of very fine limestone particles by dissolved, and later precipitated, bicarbonates. The limestone particles, suspended in runoff water, are carried down the slope. The harder "bands" would have formed subsequently in more pervious zones where soil water continues to percolate.

VIII₆: THREE-LAYER HARDENED CRUST

(Triple croûte zonaire durcie)

Ontinena, Ebro Valley, Spain

Such "layered" or "platy" zones are sometimes found in the upper portion of a crust. These very old formations are extremely hard and consist of well-crystallized calcium carbonate. The calcium carbonate of a previously formed crust has been dissolved at the surface by CO_2-charged water and recrystallized by rapid evaporation. This example shows three layered crusts that are probably of different ages.

PLATE VIII

WARM OR DRY CLIMATE ISOHUMIC SOILS

CHAPTER V

BRUNIFIED AND ELUVIATED SOILS

Noncalcareous soils occurring under temperate or cold climate, in plains or low mountains, having a mull (sometimes a moder) type of humus and a brown (B) horizon of alteration or an "argillic" B_t horizon of accumulation.

The surface of these soils consists of a well-aerated mull with crumb structure of dark brown "clay-ferric iron-humus" aggregates. The lighter colored (B) horizon of alteration has blocky structure. In this horizon, iron oxides form film coatings around the clay platelets. These are mainly micaceous minerals, either inherited or somewhat degraded through sheet-opening (illite-vermiculite).

In loose and deep sedimentary or reworked parent materials, relatively rich in inherited clay minerals, a mechanical translocation, or "eluviation," of clay and adsorbed iron is superimposed on the weathering of primary minerals (brunification). In such cases, the (B) horizon of alteration separates into a lighter colored eluvial A_2 horizon (loss of iron and clay) and a darker B_t "argillic" horizon characterized by oriented clay skins (argillans) around peds.

GENESIS OF BRUNIFIED SOILS UNDER ATLANTIC CLIMATE

Under Atlantic climate, two main avenues of evolution of Brunified soils correspond to two different parent materials.

1. Hard, often acid, crystalline or detrital rock. Weathering processes are more important and compensate for eluviation, leading either to *Acid Brown soils* when underlying parent material is poor in weatherable minerals or to *Mesotrophic* or *Eutrophic Brown soils*, in the opposite case. Soils can evolve from these parent materials by degradation of the organic matter (moder or mor), causing a more complete biochemical weathering of primary and secondary minerals. I will call this process "direct podzolization." It results in the gradual transformation of the brown, weathered "cambic" horizon into a "spodic-like" horizon by enrichment in amorphous, organic, and mineral materials. Study of the soil profiles will demonstrate the essential role played by iron acting as a "brake" to this type of genesis.

The "Acid Brown soil," which constitutes the equilibrium phase in deciduous or mixed forests (low mountains), is characterized by an "acid

69

mull" type of humus. Transformation of the mull into a moder is accompanied by a very mild beginning of podzolization along with the formation of not yet very mobile complexes of humic substances, iron, and aluminum (Bruckert et al., 1975). Iron complexes do not migrate but they contribute to the "ochre" coloring of the profile (Ochric Brown soils). As migration processes become more pronounced, Brown Podzolic soils or Podzolic soils, with more or less organic matter depending upon altitude, are formed (see Chapter VI).

2. Loose sedimentary materials. "Eluviation" of clay and iron occurs at an early stage—as soon as the soil starts to form—on non-calcareous parent material or on material having previously undergone decarbonation. The equilibrium phase, which corresponds to the deciduous forest "climax" on various parent materials, is the *Eluviated Brown soil* with an argillic B_t horizon. Under various environmental or biotic influences, eluviation may intensify and lead to acidification and depletion of the surface horizon. This leads to the *Acid Eluviated soil* with hydromorphic features at depth due to "plugging" or compaction of the B_t horizon. In such acid and poorly aerated soils, iron partially separates from clay and the two elements migrate separately. Furthermore, the top of the argillic horizon often undergoes a process of "degradation" (Jamagne, 1973; Bullock et al., 1974), whereby a loss of crystallinity with sheet-opening of the 2/1 clay minerals is accompanied by the formation of aluminous chlorite. This leads to *Degraded Eluviated soils* which often exhibit a "planosolic" or "glossic" aspect.

If degradation of the humic horizons becomes more pronounced (with formation of a moder or a mor), secondary podzolization may take place (particularly on sandy parent materials). The end-product is a "Podzolic" soil, with a "Podzolic Eluviated soil" as frequent intermediary. I will denote this kind of podzolization "indirect podzolization," as it is preceded by an intensification of the eluviation process. The spodic B horizon of the secondary Podzol is superposed on a former B_t horizon, evidence of the eluviation phase (see Chapter VI).

NOTE 1. Special mention must be made of "complex" or polycyclic (polygenetic) profiles which have developed from several superposed parent materials (often "Paleosols"):

(i) *Eluviated Brown soils or double-layer Eluviated soils*, e.g., loam on "terra fusca", loam on marl, etc. The textural differentiation resulting from eluviation is superimposed on the initial textural heterogeneity inherited from the two superposed parent materials. In the second example, the soil usually develops hydromorphic features at depth.

(ii) *Glossic Eluviated soils or Eluviated soils with fragipan.* Under Atlantic climate, these soils develop from old parent material having already undergone prolonged evolution, sometimes dating back to pre-Wurmian. The upper part of the profile, which has been reworked by cryoturbation, has undergone recent pedogenesis. Such soils are generally highly desaturated and "eluviation" is intensified. The mineral matter is much more weathered than that of modern soils. The incipient stage, the "Eluviated Brown soil", is not found. Moreover, these soils, whose deep horizons are slightly hydromorphic initially, may evolve towards a "complex Pseudogley" when a perched water table develops near the surface (see Chapter VII).

NOTE 2. *Calcic Brown soils* have been placed with the Calcimagnesian soil class for reasons of ecology and genetic affiliation. However, the *"secondary" Eutrophic Brown soil* on terra fusca of Profile IX$_4$ has been classified with Brunified soils. Its genesis is akin to that of soils formed on hard limestone (see Table 3, Chapter III).

GENESIS OF BOREAL ELUVIATED SOILS

At the southern limit of the Podzol "zone" of the boreal forest with Ericaceae (Taiga), there exists, both in North America and in the U.S.S.R., a climatic zone which is characterized by a mixed evergreen-deciduous forest giving way to deciduous forest. The intensity of "podzolization" tends to diminish and is gradually replaced by more and more pronounced "eluviation" of clay. At snowmelt, a surface water table forms above the still frozen subsoil. These temporary saturated conditions are responsible for special soil characteristics which distinguish these soils from the Eluviated soils of Western Europe developed under Atlantic temperate climate.

These characteristics appear to be tied to three simultaneous processes: (i) eluviation of clay which is the most prominent process, (ii) temporary waterlogging which induces reduction and mobilization of iron in the middle horizons, and (iii) more or less intense biochemical weathering of part of the clay minerals with formation of organomineral complexes. The third process resembles the type of "podzolization" that occurs in Western European soils, but differs from it by exhibiting more anaerobic characteristics ("hydromorphic" podzolization comparable to that which occurs in certain Podzolic Pseudogleys) (Glazovskaya, 1974; Targulian et al., 1974; Zaidelman, 1974). Moreover, this podzolization is of "climatic" type and thus does not depend on the formation of a mor.

Hence, these soils have unique features which are not yet completely known. Nevertheless, I will maintain them provisionally in the *Boreal Eluviated soil* subclass of the French soil taxonomy.

Dernovo-Podzolic soils occur under evergreen forest with birch. A thin mull or mull-moder humus layer reflects the deciduous underbush and the herbaceous plant cover of the forest floor. Eluviation and the development of a frost-induced glossic horizon are the dominant processes.

Gray Forest soils take their place to the south. They are more humic (30- to 40-cm thick A$_1$ horizon), but less acid. Paradoxically, organomineral complexes are more mobile and accumulate in B$_t$. This appears to be in contradiction with the characteristics of humic horizons which are base-saturated and biologically very active. Two seasonal genetic processes appear to oppose each other: (i) a phase of hydromorphic mobilization of organomineral complexes during snowmelt, and (ii) a phase of intense biological humification leading to the fixation of bases within the A_1 horizon during the warming and drying of the profile (Zonn and Karpachevskii, 1964).

NOTE. The Gray Wooded soils* of Northeastern Canada, developed on calcareous moraine, come closer to the Dernovo-Podzolic soils than to the Gray Forest soils of the U.S.S.R. They differ from the latter by a more marked A_0 horizon (moder) and by the frequent presence of a "calcic" horizon below the B_t horizon.

Table 6.   Brunified and Eluviated Soils: Genesis under Atlantic Climate

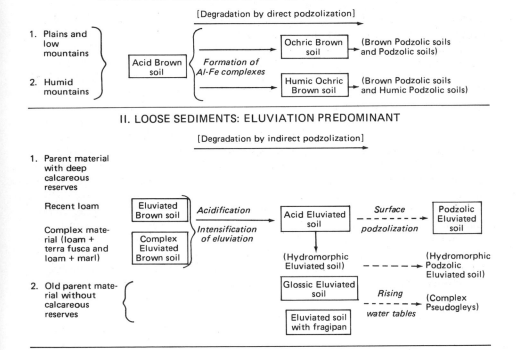

Note: The Eutrophic Brown soil on terra fusca (Profile IX_4) will be found in Table 3 (Chapter III).

*Translators' note: Since 1972, these soils are called Gray Luvisols in the System of Soil Classification for Canada (Canada Soil Survey Committee. 1973. *Proceedings 9th Meeting*, Saskatoon, Saskatchewan, p. 262).

Table 7. Eluviated Soils: Climatic Evolution on Loamy or Glacial Parent Material

Climatic Zone	Climatic Vegetation	More Continental ⟶	
Increasing mean temperature		**Atlantic** *Not very humic; nonhydromorphic at the surface*	**Continental** *More humic; hydromorphic at the surface*
Cold boreal	Evergreen forest with Ericaceae		Podzols
Transitional boreal	Mixed forest with herbaceous flora	(Acid Eluviated soil)	Dernovo-Podzolic soil
Cold temperate	Boreal deciduous forest	(Acid Eluviated soil)	Gray Forest soil
Temperate	Temperate deciduous forest	Eluviated Brown soil locally Acid Eluviated soil	(Eluviated Chernozem)[a]
Warm temperate	Deciduous forest with Mediterranean trend	(Fersialitic Eluviated Brown soil)[b]	

[a]See Profile VII$_5$ (Chapter IV).
[b]See Profile XVI$_1$ (Chapter VIII).

IX₁: ACID BROWN SOIL

(Sol brun acide)

F.A.O.: *Dystric Cambisol*, U.S.: *Typic Dystrochrept*

Location: Sainte-Marie, Bezange State Forest, Meurthe-et-Moselle, France.
Topography: Regular gentle slope (4%). Elevation 250 m.
Parent material: Rhaetian sandstone (Lower Lias).
Climate: Atlantic with continental trend. P. 720 mm; M.T. 10°C.
Vegetation: Beech grove; mull flora consists of *Melica uniflora, Lamium galeobdolon, Milium effusum.*

Profile Description

A_1 (0-10 cm):   Olive-brown (2.5 Y 4/3) sandy mull; fine crumb structure; common roots; gradual boundary.

(B) (10-60 cm):   Dark yellowish-brown (10 YR 4/4) homogeneous fine sand; very fine granular structure, aerated; common roots in first 40 cm, few roots below; clear boundary.

C: Strong brown (7.5 YR 5/6) Rhaetian sand.

Geochemical and Biochemical Properties

Particle-size distribution: *Fine* sand with 8% clay. No visible translocation of clay from top to bottom.

Exchange complex: Exchange capacity is low [3.6 meq/100 g in (B)], except in A_1 (10 meq/100 g). S/T is over 50% in A_1, but is low in the mineral horizons (8 to 14%). In A_1, there is a high concentration of exchangeable bases, particularly Ca^{2+}, due to the biogeochemical cycle. S is 5.2 meq/100 g in A_1 versus 0.3-0.4 meq/100 g in the mineral horizons. pH (water) equals 4.7 in A_1 and 3.8 in (B).

Biochemical properties: As in all acid mulls, the organic matter content is low. It decreases rapidly from 4% in A_1 to 1.2% at the top of (B). The C/N ratio is 11 in A_1, decreases rapidly in (B) to 6 at its base. This is characteristic of Brown soils. The organic matter is especially rich in nonextractable "humin of insolubilization," which is a feature of acid mulls (Toutain, 1974).

Iron and aluminum: Free iron content is quite high (0.7%) and constant throughout the profile. Aluminum shows a maximum (0.19%) at the top of (B). The aluminum/clay ratio reaches 0.023, which is a sign of incipient degradation. The complexed iron/free iron ratio is very low (0.014) which is characteristic of cambic horizons.

Genesis. The relatively high content of active free iron in the parent material restrains podzolization and leads to the pedogenic formation of a mull and to "brunification" (Toutain, 1974). On neighboring sandstone outcrops lower in iron, there is formation of a Podzolic soil with moder. Yet the composition of the litter is identical (Profile XI₂). Starting in the A_1 horizon, iron quickly insolubilizes the "phenolic polymers" resulting from the decomposition of lignin. The effect is to restrain translocation of the pseudosoluble iron and aluminum complexes. Only compounds high in saccharides and proteins remain soluble, but they undergo rapid mineralization. Insolubilized phenolic polymers make up the major portion of the humic compounds in A_1 (FA-HA-humin). They act as efficient cementing agents and promote the good structure of this horizon. Due to the insolubilization of the complexed iron in A_1, the free iron in the (B) horizon is almost totally bound to clay minerals (Bruckert et al., 1975).

REFERENCE: Toutain, 1974.

IX₂: OCHRIC BROWN SOIL

(Sol brun ocreux)

F.A.O.: *Dystric Cambisol;* U.S.: *Typic Dystrochrept*

Location: Rambervillers State Forest, Meurthe-et-Moselle, France.
Topography: Level plateau. Elevation 430 m.
Parent material: "Intermediate" sandstone (Lower Trias), reworked at the surface.
Climate: Vosges elevation zone in low mountains. P. 900 mm; M.T. 9°C.
Vegetation: Sparse oak and beech grove with scattered fir; acidophilous flora consisting of *Deschampsia flexuosa, Vaccinium myrtillus, Leucobryum glaucum.*

Profile Description

A_0A_1 (0-5 cm): Black moder; single grained, loose; contains coprogenous particles and translucent quartz grains.
A_1B_h (5-8 cm): Dark gray sand; sand grains coated with humus; massive structure; gradual boundary.
(B) (8-70 cm): Reddish-yellow (7.5 YR 6/8) sand; very fine granular structure becoming massive; common roots; gradual boundary.
IIC: Yellowish-red (5 YR 5/6) loamy sand.

Geochemical and Biochemical Properties

Particle-size distribution: Medium sand, with 9% clay in (B). No evident clay eluviation, except for some depletion in the humic horizon.

Exchange complex: Exchange capacity is low and decreases with depth [14 meq/100 g in A_1 to 3 meq/100 g in (B)]. S/T is low throughout (2.5%), except for a slight increase in A_1 (6%). pH (water) is 3.8 in A_1 and 4.6 in (B). The biogeochemical cycle of bases has less influence on the moder than on the acid mull. Nevertheless, the value of S is considerably larger in A_1 than in (B).

Biochemical properties: Organic matter decreases with depth, from 15% in A_0A_1 to 6% in A_1B_h and to 2.5% in (B), but (B) still has a fair amount of complexing organic components. C/N ratio is high throughout and increases with depth from 22 in A_1 to 33 in (B). These two characteristics reveal that podzolization is beginning under the influence of mobile organic substances.

Iron and aluminum: Free iron content varies little from top to bottom [0.30% in B_h, 0.45% in (B)]. On the other hand, migration of aluminum is quite evident as aluminum increases from 0.10% in A_1B_h to 0.26% in (B). The aluminum/clay ratio increases from 0.012 in A to 0.03 in B, indicating that degradation of clay minerals has begun. Finally, the complexed iron/free iron ratio is 0.1 in B which is indicative of a "spodic" tendency.

Genesis. Compared to the Acid Brown soil with mull, this soil shows signs of incipient biochemical podzolization, such as the characteristics of the organic matter, the mobilization of aluminum, and the high chroma of colors which is due to the formation of iron-organic matter complexes. In (B), 10% of free iron is complexed and extractable (Bruckert et al., 1975). Nevertheless, iron is not very mobile in this well-aerated environment. This is "cryptopodzolization," which is still beginning and does not lead to pronounced morphological differentiation. This intermediate stage between "brunification" and "podzolization" is a reflection of the ecological conditions: nature of the litter (moderately acidifying compared to evergreen litter), good aeration of the reworked parent material, and medium iron content (lower than in the Acid Brown soil). Bases play no role as their content is extremely low in C (0.1 meq/100 g).

REFERENCE: *Soil Map of France (1:100,000),* Saint-Dié sheet, C.N.R.S., 1973.

IX$_3$: HUMIC OCHRIC SOIL

(Sol ocreux humifère)

F.A.O.: *Humic Cambisol;* U.S.: *Umbric Dystrochrept*

Location: Black Wood, Albères Mountains, Pyrénées-Orientales, France.
Topography: Moderate slope. Elevation 800 m.
Parent material: Gneiss containing garnet, reworked at the surface.
Climate: Humid montane. P. 900 mm; M.T. about 8°C.
Vegetation: Beech forest, sparse acidophilous flora.

Profile Description

A$_0$A$_1$ (0-10 cm): Dark reddish-brown (5 YR 2/2) acid mull; crumb structure, well aerated; common fine roots.

A$_1$(B) (10-40 cm): Yellowish-brown (10 YR 5/6) less humic than above; graded texture; crumb to "fluffy" structure, friable, well aerated; many roots; gradual boundary.

(B) (40-120 cm): Yellowish-brown (10 YR 5/8); fluffy structure; much angular gravel; few coarse roots; clear boundary.

(C): Loamy sand fine earth with gneiss gravel.

Geochemical and Biochemical Properties

Particle-size distribution and clay minerals: Gravelly fine loamy sand, characteristic of gneiss. Clay eluviation, which is limited, is compensated by weathering so that the clay content (14.5%) is fairly constant to the 50-cm depth. Below that level, it decreases to 5% in C. Clays consist of micaceous minerals, mainly vermiculite.

Exchange complex: Fairly high exchange capacity [19 meq/100 g in A$_1$(B)] due to the high organic matter content and the presence of vermiculite. Base saturation is relatively low (4-6%), but increases in the humic horizons (22% in A$_1$) because of a favorable biogeochemical cycle, as is the rule in all soils with a mull (pH in water is 5).

Biochemical properties: Fresh organic matter is decomposed rapidly and humus penetrates deeply into the profile [8% O.M. in A$_1$ and 5.5% in A$_1$(B)]. C/N ratio of about 16 in A$_1$ and 14 in (B) indicates relatively intense biological activity.

Iron and aluminum: Free iron and aluminum contents are quite high with 1.5 and 0.26%, respectively, in A$_1$(B). Like clay, they are fairly constant in the upper 50 cm of the profile, as leaching is compensated by weathering. The aluminum/clay ratio in the (B) horizon is 0.023. It shows that primary minerals and clays are being weathered, which is characteristic of "Ochric" soils.

Genesis. Very similar to the previous soil, this soil is nevertheless more humic due to a more pronounced montane climate. It can be viewed as an "analogue" of "Humic Brown soils" which are found at the humid montane elevation zone, under deciduous or mixed forest cover and on widely different parent materials (climatic climax). Here, chemical podzolization is slightly more pronounced than in the previous soil. It is revealed by the ocherous color of the (B) horizon, its "fluffy" structure, and an aluminum/clay ratio above 0.02. This moderate podzolization is not linked to the formation of a "moder" type of humus as in Ochric soils in plains. It is characteristic of the humid montane climate and even occurs in the presence of a "mull" ("climatic" podzolization). On very exposed summits with subalpine meadow, this soil gives way to Humic Cryptopodzolic Rankers (Profile IV$_1$).

REFERENCE: *Soil Map of France (1:100,000),* Perpignan sheet, Soils Laboratory, Montpellier.

IX$_4$: EUTROPHIC BROWN SOIL (on terra fusca)

(Sol brun eutrophe)

F.A.O.: *Chromic Cambisol;* U.S.: *Lithic Eutrochrept*

Location: "Bois Fourasse," Liverdun, Massif de Haye, Meurthe-et-Moselle, France.
Topography: Plateau. Elevation 310 m.
Parent material: Terra fusca and hardened Bajocian limestone, breaking into plates.
Climate: Atlantic with continental trend. P. 740 mm; M.T. 9.5°C.
Vegetation: Oak forest with hornbeam changing into shrubs. Many calcicole shrubs consisting of *Lonicera xylosteum, Melica nutans, Melica uniflora,* and *Asarum europaeum.*

Profile Description

A$_1$ (0-6 cm):   Brown (7.5 YR 5/3) mull; angular crumb structure, well aerated; common fine and medium roots.
(B) (6-23 cm):   Strong brown (7.5 YR 5/6) silty clay terra fusca; fine blocky structure; plastic when wet, porous; common subhorizontal roots; not effervescent; abrupt transition.
R: Bajocian limestone slab, fractures into horizontal plates; fine earth is effervescent within 0.5 cm of contact with limestone.

Geochemical and Biochemical Properties

Particle-size distribution and sesquioxides: Silty clay material with 41% clay and 2% sand, very rich in iron oxides (4% free iron) and rich in aluminum (0.8% free aluminum). These values are characteristic of *terra fusca* when it has not (or only slightly) been contaminated with external elements. Carbonates are absent in the fine earth, except in the immediate vicinity of the lithic contact.

Exchange complex: Exchange capacity is relatively low [26 meq/100 g in (B)] considering the high contents of humus and clay. This is due to the presence of kaolinite in the terra fusca. The exchange complex is base-saturated in (B) and slightly desaturated in A$_1$ (80%; pH in water is 5.8).

Biochemistry: Active forest mull with high organic matter content [8% in A$_1$ and 5% in (B)] due to the high content of insolubilizing agents (iron and clay) and to the dense plant cover which provides an abundant litter. The color of the organic matter is lighter than that of carbonaceous mulls because polymerization is lower. C/N ratio is 16 in A$_1$ and 14 below.

Genesis. Eutrophic Brown soils may have two "genetic" origins. Some are "monocyclic." They result from the direct weathering of base-rich rock (pelite, schist, diorite, etc.). Their genesis is comparable to that of Acid Brown soils but they contain more exchangeable bases. Others are "polycyclic," as is this soil. These result from the "brunification" of a Paleosol which, in the present case, is a Fersialitic soil—"terra fusca." Under the action of various periglacial processes (cryoturbation, erosion, etc.), the evolution of terra fusca will be controlled by its thickness and the nature of the underlying limestone. Polycyclic Brunified Rendzinas form on finely divided, soft limestone (Profile VI$_1$), whereas on hard platy limestone, terra fusca becomes "brunified" and loses its reddish color because of a change of state of iron oxides. There is no "recarbonation" and a slightly desaturated forest mull forms at the soil surface. In fact, this is an "analogous soil" under deciduous forest with mull. The present profile is exceptionally shallow; it is normally thicker (see Profile II$_1$).

REFERENCE: Gury, M., *Soil Map of the Plateau of Haye,* C.N.R.S., Nancy, 1972. (Photo by M. Gury.)

IX₅: ELUVIATED BROWN SOIL

(Sol brun lessivé)

F.A.O.: *Orthic Luvisol;* U.S.: *Typic Hapludalf*

Location: Cilly, northeast of Laon, Aisne, France.
Topography: Level plateau. Elevation 135 m.
Parent material: Recent loess (Wurm III).
Climate: Temperate, moderately Atlantic. P. 680 mm; M.T. 9.7°C.
Vegetation: Ruderal elm-grove copse.

Profile Description

A₁ (0-16 cm):	Dark gray-brown (10 YR 4/2) mull, becoming lighter with depth; crumb structure, friable; common fine roots; gradual boundary.
A₂ (16-36 cm):	Pale brown (10 YR 6/3) silt loam; fairly compact, massive structure; gradual boundary.
A/B (36-55 cm):	Darker than above; beginning of clay accumulation.
B_t (55-120 cm):	Yellowish-brown (10 YR 5/6) argillic horizon; moderate blocky structure with dark brown (7.5 YR 4/4) clay coatings, tends towards prismatic structure at the base; diffuse boundary.
C (120-200 cm):	Yellowish-brown (10 YR 5/6) silt loam; massive structure.

Geochemical and Biochemical Properties

Particle-size distribution, iron, and clay: Homogeneous windblown silt loam, very low in sand (1%) and fairly high in clay (19%). Eluviation of clay and of bound iron results in *minimum clay and iron contents* in the A horizons (14.5% clay, 0.9% iron in A₂) and a maximum in B_t (27.5% clay, 1.7% iron). The translocation indices of both constituents have values of the same order of magnitude (1/9); losses from the A horizons approximately balance gains in the B horizon. *Clay minerals* are "inherited," only slightly degraded, and of the 2/1 type (illite, vermiculite) (Jamagne, 1973).

Exchange complex: The value of S is lowest in the A₂ and A/B horizons (4.3 meq/100 g), increases in B_t (8.7 meq/100 g) but mostly in A₁ (12 meq/100 g), indicating an efficient biogeochemical cycle and a good retention of bases in the mull. Therefore, the humus is almost base-saturated at the surface (pH 6.7). On the contrary, S/T is at a minimum, at about 50%, in A/B (pH 4.8); it increases slightly in B_t.

Biochemical properties: "Active" mull with 4.3% organic matter at the surface, decreasing very gradually with depth. C/N ratio of 12 in A₁, decreasing to below 8 in B_t, where the organic matter content is very low (less than 1%—property of argillic horizons).

Physical properties: Well-aerated profile, porosity greater than 50% in A₁, 49% in A₂, and still 42% in B_t.

Genesis. In loose and homogeneous parent materials, brunification is associated with "eluviation" or "mechanical" translocation of the fine clay particles together with their iron oxide coatings. The B_t horizon is an "argillic" horizon, low in organic matter which is completely mineralized in the A horizons, and plays no role in the translocation process. This monocyclic type of genesis is observed in windblown loams, or in the most recent moraines, still often calcareous at depth. Eluviation takes place as soon as decarbonation of the surface horizons is terminated (Schwertmann, 1968). The biogeochemical cycle of bases is much more efficient in Eluviated Brown soils with mull than in Eluviated soils containing an acid moder. This latter type of organic matter is characterized by slow humification and retains bases less well.

REFERENCE: *Soil Map of France (1:100,000),* Laon sheet, Chambre d'Agriculture de l'Aisne, 1973.

IX₆: COMPLEX ELUVIATED BROWN SOIL

(Sol brun lessivé complexe)

F.A.O.: *Chromic Luvisol;* U.S.: *Typic Hapludalf*

Location: Sivrite Arboretum, Massif de Haye, Meurthe-et-Moselle, France.
Topography: Level site, at bottom of slight depression in limestone plateau.
Parent material: Eolian loam and reworked terra fusca, resting on Bajocian limestone slab.
Climate: Atlantic with continental trend. P. 730 mm; M.T. 9.5°C.
Vegetation: Beech, hornbeam, and robur oak forest; fresh mull flora consists of *Asperula odorata, Lamium galeobdolon, Deschampsia coespitosa.*

Profile Description

A_1 (0-8 cm):   Dark brown (7.5 YR 3/2) mull; crumb structure, aerated; common horizontal roots; gradual boundary.

A_2 (8-40 cm):   Brown (7.5 YR 5/4) loam; fine crumb to subangular blocky structure, aerated; few roots, intense biological activity; humic spots at places; clear boundary.

B_t (40-70 cm):   Strong brown (7.5 YR 5/6) clay loam; strong blocky structure, with ocher clay skins; firm, compact, slightly porous; few roots; clear boundary.

BC_g (70-100 cm):   Yellowish-red (5 YR 4/6) clay; greenish mottles; coarse blocky to massive structure; resting on limestone slab.

Geochemical and Biochemical Properties

Particle-size distribution: Loam containing 20-25% clay and about 50% silt in the upper 40 cm. The B_t and BC_g horizons are enriched with clay, with 39 and 44%, respectively. The translocation index for clay is close to ½. However, this value is not significant due to the complexity of the parent material.

Exchange complex: Moderate exchange capacity (11 meq/100 g in A_2), typical of micaceous clay minerals. S/T displays two maxima: one in A_1 (75% with S at 12 meq/100 g) and another in the B_t and BC_g horizons (65 and 80%, respectively). S/T is lowest in A_2 (32%). pH (water) is 6.4 in A_1, 5 in A_2, and 5.4 in B_t. These values indicate the role of the biogeochemical cycle in A_1 and the moderate accumulation of bases in the B horizons.

Biochemistry: Mesotrophic mull. Formation of slightly polymerized and rapidly biodegraded humic substances in A_1 (brown HA). Only 5% organic matter in A_1. C/N ratio decreases from 15 in A_1 to 10 in B_t. Such values are characteristic of Eluviated Brown soils with mull.

Iron and aluminum hydroxides: Their content is high in B_t and especially in BC_g due to the presence of *terra fusca* (1.6% iron and 0.66% aluminum). The distribution of hydroxides is similar to that of clay in A_2 and in the B horizons.

Genesis. The microstructure of the B_t horizon reveals the complexity of the profile (see Appendix: Micrographs XX_8 and XX_9). Yellow ferri-argillans in place can be observed next to disturbed and disrupted reddish argillans (papules) formed from terra fusca. Thus, below a surface layer of undisturbed loam, there is a mixed soliflucted layer (loam and terra fusca) which has been subsequently enriched with illuviated clay (B_t horizon). Terra fusca (with hydromorphic mottles at places) predominates in BC_g. This soil is called "Eluviated Brown soil" by analogy with homogeneous Eluviated Brown soils. Their properties—humus type, porosity, color, structure, exchange complex status, moderate eluviation of clay and iron—are similar. It may be inferred from this that their overall genesis is also comparable.

REFERENCE: Gury, M., *Legend of Soil Map of the Plateau of Haye,* C.N.R.S., Nancy, 1972. (Photo by M. Gury.)

PLATE IX

ATLANTIC BROWN SOILS AND ELUVIATED BROWN SOILS

X₁: HYDROMORPHIC ACID ELUVIATED SOIL

(Sol lessivé acide hydromorphe)

F.A.O.: *Orthic Luvisol;* U.S.: *Aquic Hapludalf*

Location: Dommartin-lès-Toul Forest, Meurthe-et-Moselle, France.
Topography: Gentle slope, northwest exposure. Elevation 280 m.
Parent material: Loam and *old* alluvium overlying Bathonian marl.
Climate: Atlantic with continental trend. P. 740 mm; M.T. 9.4°C.
Vegetation: Mixed beech-oak forest with hornbeam; *Lonicera periclymenum, Deschampsia coespitosa, Hedera helix, Polytrichum formosum.*

Profile Description

A_0A_1 (0-5 cm):	Dark grayish-brown (10 YR 4/2) weakly structured mull-moder, bare quartz grains; clear boundary.
A_2 (5-30 cm):	Very pale brown (10 YR 8/4) sandy loam; single-grained to massive structure, fairly porous; common roots.
A/B (30-45 cm):	Gradual enrichment with clay.
B_t (45-70 cm):	Reddish-yellow (7.5 YR 6/6) clay loam, with small brightly colored spots; fine blocky structure with clay coatings on peds; few fine roots; gradual boundary.
IIB_g (70-140 cm):	Brownish-yellow (10 YR 6/6) decarbonated marl, with alternating gray and elongated rust mottles; weak blocky structure; gradual boundary to calcareous marl at 140 cm. (*Note:* This horizon is not visible in the photograph.)

Geochemical and Biochemical Properties

Particle-size distribution: Particle-size summation curves show that the soliflucted material in the first 70 cm is homogeneous. Below this depth, decarbonated Bathonian marl, with 53% clay, is present. The clay accumulation in B_t can be attributed solely to illuviation as the translocation index is close to 1/3 (12.6% clay in A_2). Two kinds of clay minerals, vermiculite and inherited kaolinite, are found, but only vermiculite migrates.

Exchange complex: Exchange capacity is high in the marl. It is minimum in A_2. S/T is also lowest in A_2 (5.4%) and increases to 44 and 30% in A_0A_1 and B_t, respectively. The decarbonated marl is almost base-saturated. Extreme acidity is strongest in A_2 (pH values of 4.1 in A_2, 4.4 in A_0A_1, and 4.6 in B_t).

Biochemical properties: Organic matter is 9.5% in A_1 and decreases very rapidly in A_2B and B_t (0.6 and 0.4%, respectively). C/N ratio is 15.5 in A_1 and decreases rapidly with depth to 6 in B_t. This is characteristic of nonpodzolized Eluviated soils.

Iron and aluminum: The translocation index for iron is half that for clay. Free aluminum content is low, but the aluminum/clay ratio decreases from 0.023 at the surface to 0.01 at depth (slight weathering at the surface).

Micromorphology: In the B_t horizon, iron-rich ocher argillans alternate with partly bleached argillans (tendency towards separate migration of clay and iron).

Genesis. The profile has two layers: the first is 70 cm thick and consists of a mixture of eolian loam and sand from terraces which have been homogenized by solifluction; the second is formed of decarbonated marl. The upper portion of the profile constitutes a typical *Acid Eluviated soil*, whereas the marl has acquired marked hydromorphic features—hence the designation *Hydromorphic Acid Eluviated soil*. Because of the nature and the age of the parent material, this soil is more developed than an Eluviated Brown soil because of acidification of the humus, a slight degradation of structure, a decrease in porosity, and more intense "eluviation" of clay and iron inducing a partial bleaching of the surface horizons. Nevertheless, signs of degradation are not yet very distinct.

REFERENCE: Gury, M., *Legend, Soil Map of the Plateau of Haye,* C.N.R.S., Nancy, 1972.

X₂: PODZOLIC ELUVIATED SOIL

(Sol lessivé podzolique)

F.A.O.: *Humic Podzol;* U.S.: *Alfic Haplorthod*

Location: La Tillaie, Fontainebleau Forest, Seine-et-Marne, France.
Topography: Plateau, very gentle slope. Elevation 137 m.
Parent material: Stampian sand "blown" over Beauce limestone.
Climate: Moderate Atlantic. P. 700 mm; M.T. 9°C.
Vegetation: Acidophilous beech grove; *Pteridium aquilinum, Carex pilulifera.*

Profile Description

A_0A_1 (0-8 cm):   Very dark gray (10 YR 3/1) moder with surface organic horizon; single-grained structure; transparent quartz grains.

A'_2 (8-15 cm):   Light gray (10 YR 7/2) bleached fine sand; single-grained.

B_h (15-35 cm):   Dark brown to brown (10 YR 4/3) fine sand, gradually lighter with depth; quartz grains coated in brown.

A_2 (35-75 cm):   Pale brown (10 YR 6/3) fine sand, darker with depth; single-grained structure; gradual boundary.

β (75-90 cm):   Strong brown (7.5 YR 5/6) sandy loam; moderate blocky structure with ocher coatings; horizontal wavy rust bands.

IIC: Sand with calcareous gravel; effervescent.

Geochemical and Biochemical Properties

Particle-size distribution: Very fine sand, analogous to undisturbed Stampian sand but slightly higher in silt (8-10%). Clay content is nil or very low, except in β (17.2%). The spodic-like B_h horizon contains from 0.5 to 1% clay.

Exchange complex: Exchange capacity is very low throughout. So is the value of S (except in β and C). However, the A_1 horizon contains 1.2 meq/100 g of exchangeable bases or three times as much as in the underlying horizon. Base saturation increases from 16% in A'_2 and B_h (pH in water is 4.2) to 40-45% in A_2 (pH in water is 5.2). The β horizon is base-saturated.

Biochemistry: The moder contains 5% organic matter with a C/N ratio of 20. With 1% organic matter and a C/N ratio of 22, the B_h horizon shows a slight penetration of organic matter mainly in the form of extractable FA and HA. Below this depth, both organic matter and the C/N ratio decrease abruptly.

Iron and aluminum: The β horizon is rich in free iron (1.36%), whereas the remainder of the profile contains little iron. There is a slight accumulation in B_h (0.30% as opposed to 0.14% in A'_2). A marked sign of podzolization can be found in the free aluminum/clay ratio which increases from a low value of 0.0025 in B_t to 0.07 in B_h.

Genesis. This soil displays two imbricated profiles: a weakly developed podzolic profile at the surface and, at depth, an eluviated profile with a β horizon enriched in iron and clay and resting on calcareous sand. In fact, the β horizon appears to have a mixed origin. It contains montmorillonite from decarbonation of the calcareous parent material and also vermiculite which results from eluviation of the eolian sands and predominates in the upper part of this horizon (Robin and De Coninck, 1975). The upper podzolic profile could have resulted from a degradation caused by a change in vegetation. This does not appear to be the case here. "Eluviation" of the surface sand has probably resulted in a profile depleted of clay, iron, and calcium to the extent that a moder of deciduous leaves could form and that podzolization could begin at the surface (see Profile XI_5, Humic Podzol).

REFERENCE: Robin, 1968.

X₃: HYDROMORPHIC GLOSSIC ELUVIATED SOIL
(Sol lessivé glossique hydromorphe)

F.A.O.: *Gleyic Acrisol;* **U.S.:** *Glossic Fragiudult*

Location: "La Croix du Soldat," Sainte-Hélène Forest, Meurthe-et-Moselle, France.
Topography: Level site. Elevation 340 m.
Parent material: Old loam (Riss ?), 4 m thick, resting on alluvial terrace.
Climate: Atlantic with continental trend. P. 900 mm; M.T. 9°C.
Vegetation: Acidophilous beech and oak forest; *Luzula albida, Deschampsia flexuosa.*

Profile Description

A_0A_1 (0-10 cm): Black moder, becoming less humic and paler with depth; fine crumb to single-grained structure.
A_2 (10-25 cm): Yellow (2.5 Y 7/6) loam; massive structure; many roots; gradual boundary.
A_{2g} (25-45 cm): Pale yellow (2.5 Y 7/4) loam with yellow (10 YR 7/6) or light gray (2.5 Y 7/2) diffuse patches; massive structure; abrupt boundary.
B_g (45-108 cm): Brownish-yellow (10 YR 6/6) clay loam with many bleached vertical bands ("tongues"); coarse blocky structure with clay coatings; many iron-manganese concretions (the glossic aspect grades into a fragipan at depth).
B/C: Reddish yellow (7.5 YR 6/8) clay loam.

Geochemical and Biochemical Properties

Particle-size distribution: Pronounced depletion of clay in the A_2 and A_{2g} horizons; clay accumulation remains dispersed throughout B_g (12.5% in A_{2g}, 33% in B_g, and 31% in B/C). Particle-size summation curves indicate that the material is *homogeneous* from A_1 to B/C.

Exchange complex: Exchange capacity and S/T are minimum in A_2 (S/T is 14% in A_1, 11% in A_2, 21% in B_g, and 33% in B/C). pH (water) ranges from 4 in A_1 to 5 in B_g.

Biochemical properties: Humus forms a moder with low biological activity (29% O.M. in A_1 with a C/N ratio of 22.3).

Iron and aluminum: Migration of free iron is comparable to that of clay (0.7% in A_{2g} and 1.8% in B_g). Free aluminum (0.30%) is quite uniformly distributed. The aluminum/clay ratio decreases from A to B but does not exceed 0.023 in the A horizons. This indicates a slight weathering in the A horizons.

Micromorphology: (i) *At the base of A_{2g} and within the upper portion of B_g,* two kinds of argillans are found: weakly expressed and lightly colored argillans in place; and fragments of ferri-argillans which are thick, broken, and reincorporated into the matrix (papules). (ii) *Along the "tongues,"* presence of thick, but undisturbed, ferri-argillans, which, in the center, give way to argillans composed of deferrified and poorly crystallized clay minerals (vermiculites and aluminous chlorites—a sign of *weathering*).

Genesis. This is a complex and polycyclic Eluviated soil. The B_g and B/C horizons with their brownish-yellow, often slightly rubefied, matrix color can be considered to represent a very old "Plastosol" (high proportion of kaolinite coexistent with vermiculite). The formation of "tongues" is a periglacial phenomenon: shrinkage cracks in the Plastosol were filled with eluviated clay and iron. More recently, acid ground water flowing preferentially within these "tongues" has deferrified, bleached, and even weathered the fine clay particles. The upper portion of the profile (A_2 and A_{2g}) has been reworked by cryoturbation. The B_t horizon, having become partially impervious, has enabled a very temporary water table to form in A_{2g} (yellow mottles). Such a soil could just as well have been designated Eluviated soil with pseudogley.

REFERENCES: Le Tacon, 1966; *Soil Map of France (1:100,000),* Saint-Dié sheet, C.N.R.S., Nancy, 1973.

X₄: HYDROMORPHIC ELUVIATED SOIL WITH FRAGIPAN

(Sol lessivé hydromorphe à fragipan)

F.A.O.: *Gleyic Luvisol;* U.S.: *Aquic Fragiudalf*

Location: Saint-Gobain State Forest, Aisne, France.
Topography: Plateau with gentle slope. Elevation 195 m.
Parent material: Old loam (Wurm I ?).
Climate: Temperate, moderately Atlantic. P. 680 mm; M.T. 9.7°C.
Vegetation: Oak forest with beech.

Profile Description

A_1 (0-10 cm):	Dark brown (10 YR 3/3) acid mull; fine crumb structure; many roots, clear wavy boundary.
A_2 (10-25 cm):	Light gray (10 YR 7/2) silt loam, fairly compact; fine blocky structure; few roots.
A_{2g} (25-45 cm):	Same as above but presence of many yellowish-red (5 YR 4/6) or light gray (5 Y 7/1) hydromorphic mottles; clear boundary.
B_{tg} (45-80 cm):	Yellowish-brown (10 YR 5/6) clay loam, yellowish-red (5 YR 4/6) and fine, reduced, light gray (5 Y 7/1) mottles; massive to fine blocky structure, brown coatings; few vertical roots.
C_{gx} (80-130 cm):	Strong brown (7.5 YR 5/8) silt loam "fragipan"; hard, dense massive structure; few yellowish-red mottles; vertical cracks filled with light gray (5 Y 7/1) clay.

Geochemical and Biochemical Properties

Particle-size distribution: Silt loam with 20-22% sand throughout the profile. Clay is eluviated from A (10% clay) to the top of B_t (close to 30%), then clay content decreases gradually to 20% in the fragipan. Dominant clay minerals are aluminous vermiculite, which is a sign of moderate weathering (Jamagne, 1973).

Exchange complex: Like in all Eluviated soils, exchangeable bases, as well as S/T, display two maxima (one in A_1 under the influence of the biological cycle, the other in B_{tg} and C_{gx}) and a minimum in A_2. S decreases from 6 meq/100 g in A_1 to 1 meq/100 g in A_2 (S/T less than 25%), then increases to 5 meq/100 g in B_{tg} (S/T 50%). pH (water) varies little from 4.6 in A_1 to 5 in B_{tg} and C_{gx}.

Biochemistry: Acid mull with a moder trend. Organic matter content is 9% at the top of A_1 and 4% at its base, with a C/N ratio of 14.

Iron hydroxides: The high free iron content is a sign of a long evolution. There is a marked depletion in A (0.7%), but from B_{tg} to C_{gx} iron content remains constant at 2%.

Genesis. This profile has undergone the same type of complex genesis as the previous soil. The loamy fragipan horizon is very dense and poorly aerated and may be considered as a "relic" formation. Eluviation of clay and free iron started long ago when the medium was still aerated (brown ferri-argillans). The fine clay particles that have migrated deeply into the shrinkage cracks of the fragipan have been subsequently deferrified and weathered by acid ground water. A portion of the eluviated clay has "plugged" the B_{tg} horizon, which has since probably been reworked. Consequently, hydromorphism, accompanied by iron segregation, has been accentuated in the A_{2g} and B_{tg} horizons.

REFERENCE: *Soil Map of France (1:100,000),* Laon sheet, Laon Agricultural Research Station, 1973.

X$_5$: DERNOVO-PODZOLIC SOIL (Boreal Glossic Eluviated soil)
(Sol dernovo-podzolique) (sol lessivé glossique boréal)

F.A.O.: *Gleyic Podzoluvisol;* **U.S.:** *Eutric Glossoboralf*

Location: 107 km northeast of Moscow, along the Moscow to Iaroslav road, U.S.S.R.

Topography: Level site in gently undulating area. Elevation 220 m.

Parent material: Windblown loam (2 m thick) resting on moraine deposits.

Climate: Cold continental. P. 550 mm; M.T. 5°C (Jan. −12°C, July +20°C).

Vegetation: Mixed boreal forest of spruce and birch. Deciduous underbrush with maple, mountain-ash and hazel. Herbaceous mull flora with *Asperula* and *Lamium*.

Profile Description

A$_0$A$_1$ (0-8 cm):	Dark grayish-brown (10 YR 4/2) to grayish-brown (10 YR 5/2) mull or mull-moder at places; crumb structure; many roots; clear wavy boundary.
A$_1$A$_2$ (8-25 cm):	Pale brown (10 YR 6/3) silt loam; very fine granular to platy structure; gradual boundary.
A$_{2g}$ (25-45 cm):	Light gray (2.5 Y 7/2) sandy loam; loose, tendency towards platy structure; abrupt boundary.
A/B (45-65 cm):	Beginning of irregular, funnel-shaped, white sandy loam "tongues" alternating with compact masses of brown loam with subangular blocky structure; small black concretions; gradual boundary.
B$_{tg}$ (65-200 cm):	Yellowish-brown (10 YR 5/6) clay loam, compact; prismatic structure; narrow vertical tongues filled with white (10 YR 8/1) powdery loam and brown cutans on their external surfaces; abrupt irregular boundary with moraine material.

Geochemical and Biochemical Properties

Particle-size distribution: Loamy parent material with 50 to 55% silt. Visible clay migration; 11% clay in A$_1$A$_2$, 7% in A$_{2g}$, 20% in A/B, 33% in B$_{tg}$, then very gradual decrease with depth (deep and dispersed penetration). Clay fraction is a mixture of kaolinite, illite, and interstratified minerals. Kaolinite does not migrate and therefore shows a relative accumulation in A$_2$ as illite migrates to B$_t$. In the A horizons, interstratified minerals disappear completely and are replaced by secondary aluminous vermiculite.

Exchange complex: Moderate exchange capacity which is minimum in A$_{2g}$ (8 meq/100 g) and maximum in B$_{tg}$ (26-28 meq/100 g). S/T is 75% in A$_1$, decreases in A$_{2g}$ (pH 5), then increases again to 80% in B$_{tg}$ (pH 5.4).

Biochemistry: Mull with rapid decomposition and a C/N ratio of 12 which decreases in A$_2$ and B. Organic matter content decreases abruptly from 6% in A$_1$ to 0.4% in A$_2$ and 0.3% in B$_{tg}$. FA/HA ratio close to 1. Cutans on the external surfaces of tongues in B$_{tg}$ are high in extractable organic matter (1.8%).

Hydroxides: The distribution of free iron closely follows that of clay. From a minimum in A$_2$ (0.5%), free iron increases in B$_{tg}$ (1.4%), then decreases slowly. Free aluminum varies little throughout the profile. Within *cutans*, iron amounts to 2% and aluminum to 0.6%, or 2% of clay content.

Genesis. This soil has common features with the Glossic Eluviated soil of the Atlantic temperate regions: clay eluviation is clearly apparent but its accumulation is dispersed and extends over a considerable depth. At snowmelt, the temporary waterlogged conditions in A$_{2g}$ produce a large elimination of iron. The frost-induced "glossic" feature, which, under Atlantic climate, is inherited from glacial phases, is in perfect correlation with the boreal climate. According to Targulian et al. (1974) and Glazovskaya (1974), the occurrence of hydromorphic "podzolization" is demonstrated by the clay balance which shows a loss of 500 kg/ha and by the presence of iron-aluminum-humus complexes within the argillans of the "tongues." It must be emphasized, however, that these organomineral complexes are neither morphologically, nor analytically, easily detected. This kind of soil is characteristic of the boreal zone below the "Podzol" zone. Moderate podzolization is not tied to localized degradation of a moder or mor humus as is the case under Atlantic climate. Rather, it constitutes a generalized bioclimatic process.

REFERENCE: Targulian, V.O., A.G. Birina, A.V. Kulikov, T.A. Sokolova, and L.K. Tselischcheva, *Guidebook to the Dernovo-Podzolic soil, 10th International Congress of Soil Science,* Moscow, 1974.

X₆: GRAY FOREST SOIL
(Sol gris forestier)

F.A.O.: *Orthic Greyzem;* U.S.: *Typic Argiboroll*

Location: Zhiguli Natural Reserve, U.S.S.R.
Topography: Moderate slope, south exposure. Elevation 380 m.
Parent material: Reworked Carboniferous marl.
Climate: Subhumid continental. P. 600 mm; M.T. 4°C (Jan. -13°C, July $+20$°C).
Vegetation: Deciduous forest with *Tilia cordata, Acer platanoides, Quercus pedunculata, Populus tremula,* and *Coryllus avellana.* Active mull herbaceous flora.

Profile Description

A_1 (0-35 cm):   Dark grayish-brown (10 YR 4/2) clay loam; well-aerated crumb structure, becoming coarser with depth; common roots; gradual boundary.

A_2 (35-50 cm):   Light gray (10 YR 7/1) loam; blocky to subangular blocky structure, white powdery coatings on peds; abrupt boundary.

B_t (50-80 cm):   Yellowish-brown (10 YR 5/6) clay; blocky structure, dark yellowish-brown (10 YR 3/4) clay-humus coatings on peds often forming vertical streaks; effervescent; abrupt boundary.

BC: Ocher, platy weathering product of Carboniferous marl; small white and friable concretions; strongly effervescent.

Geochemical and Biochemical Properties

Particle-size distribution and clay minerals: Parent material is a calcareous loam with 21% clay and 60% $CaCO_3$. Clay migration is very pronounced (28% in A_1, 22% in A_2, and 50% in B_t). The original clay minerals consist of illite and montmorillonite. Montmorillonite migrates preferentially. The minimum clay content in A_2 suggests that a portion of the clay minerals in this horizon are possibly being weathered.

Exchange complex: High exchange capacity (about 30 meq/100 g) in A_1 and in B_t due to the nature of the clay minerals. The A_1 horizon is almost base-saturated with Ca^{2+} and accessorily with Mg^{2+} (pH 6.7). A_2 is partially desaturated (pH 5.2). S/T is 100% in the slightly calcareous B_t horizon.

Biochemistry: The A_1 horizon contains 6 to 7% organic matter. Gray, polymerized humic acids predominate. The proportion of fulvic acids, especially mobile forms, increases in B_t where the FA/HA ratio reaches 3. C/N ratio is fairly high (19 in A_1 and 23 in B_t). These figures suggest some degree of podzolization.

Hydroxides: The distribution of free iron is comparable to that of clay. There is a marked migration of free aluminum with a translocation index of 1/10 between A_1 and B_t.

Genesis. The Gray Forest soil is a "zonal" soil characteristic of the deciduous forest in the U.S.S.R. It occupies an intermediate position between Dernovo-Podzolic soils to the north and Eluviated Chernozems to the south. The genesis of this soil is very complex. The mechanical translocation of fine clay particles (montmorillonite) and their deposition as argillans in the B_t horizon appears evident. But the genesis of this profile is further complicated by other processes peculiar to cold continental climates. Although evident, the strong humification and "maturation" of the organic matter in the A_1 horizon are not as pronounced as in Chernozems. Furthermore, during thawing, the base of the A_1 horizon becomes saturated by a temporary surface water table. Under anaerobiosis, the less stable humic compounds would be brought into solution and would form mobile organomineral complexes likely to migrate and accumulate as characteristic black coatings in the B_t horizon. This process is close to "podzolization" insofar as it is accompanied by a certain degradation of clay minerals in the more acid A_2 horizon (Zaidelman, 1974). Morphological and analytical criteria tend to support this assumption. This type of "hydromorphic" podzolization differs from the podzolization occurring in well-aerated sandy materials.

REFERENCE: Samoilova, *Volga-Don field trip, 10th International Congress of Soil Science,* Moscow, 1974.

PLATE X

ACID ELUVIATED SOILS AND BOREAL ELUVIATED SOILS

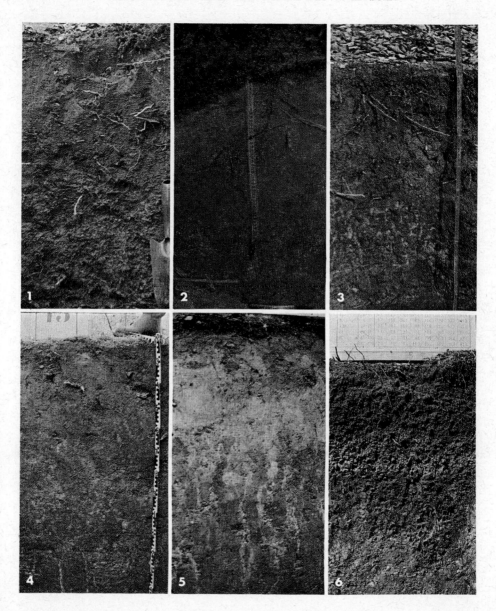

CHAPTER VI

PODZOLIZED SOILS

Soils that are characterized by the biochemical weathering of silicates by soluble and acid organic matter, with formation of more or less mobile organomineral complexes which accumulate in one or two "spodic" type horizons rich in amorphous material (dark B_h horizon enriched with humic acids and ocherous B_s horizon enriched with hydroxides).

Generally, the formation of spodic horizons is related to the presence of other characteristic horizons, such as (i) a mor or an acid moder at the surface with slow decaying rate and (ii) an *ashy* A_2 horizon ("Podzol" comes from the Russian name *pod*, which means ashes) comprised of residual silica after the mobile iron and aluminum complexes have been leached.

However, the genetic relationship between the "spodic" horizon of accumulation and the presence of A_0 (mor) and A_2 (ashy) horizons is not necessarily constant. In some profiles, the spodic horizon forms near the surface and the A_2 horizon is missing (Brown Podzolic soils). Moreover, under some humid or boreal montane climates, podzolization may occur even under a mull (climatic podzolization; see Profile IV_1, Cryptopodzolic Ranker).

It seems useful here to recall briefly the nomenclature of the French taxonomy: the *Brown Podzolic soil* has no well-differentiated A_2 horizon (only ashy spots or a thin ashy band), the brown B_h horizon is not clearly distinct from the B_s horizon which is never cemented. In *Podzolic soils*, the A_2 horizon is not completely ashy and, as in the former soil, the accumulation of humic substances in the B horizon is still small and not very conspicuous. (This soil is difficult to distinguish from the Ferruginous Podzol described by some authors, which differs from the Podzolic soil only by its more ashy A_2 horizon.)

In these transitional soils, the complexed iron/free iron ratio in the B horizon exceeds 0.25 (Bruckert et al., 1975). This clearly separates this type of "spodic" horizon from a cambic horizon.

The profiles of *Podzols* are more differentiated. The A_2 horizon is completely ashy, whereas the "spodic" horizon of accumulation is generally clearly divided into a black B_h horizon and a rust B_s horizon (which is often cemented into an "alios"). This is the Humo-Ferric Podzol. When the iron content is low, however, the B_s horizon is not present (Humic Podzol).

Genesis. The genesis of Podzolized soils is influenced by three basic ecological factors: (i) climatic factors; (ii) nature of parent material (always acid); and (iii) local "drainage" conditions. The general bioclimatic factors will receive special attention because of their great importance. The ecological significance of Podzolized soils varies greatly with vegetation "zones" or "elevation zones." Under boreal or alpine (and subalpine) climate, Podzols are "climatic" soils in equilibrium with climate and vegetation and are found on various types of parent material. Thus, these are "primary" Podzols. On the contrary, under tropical climate, Podzols are always very localized and occur only under very special site conditions. They are generally hydromorphic and associated with the presence of a "water table" on sandy material. These are "site climaxes." Finally, under Atlantic climate, the ecology of Podzols is more complicated. The nature of the parent material (always acid and low in calcium) is of paramount importance. Podzolized soils occur only on quartz sands that are low in clay and weatherable minerals, and therefore in *free iron*. *Brown Podzolic soils* and *Podzolic soils*, which are immature, are still in equilibrium with a forest vegetation close to its climax (sometimes with clearings and partially degraded) that consists of either deciduous trees in plains or mixed evergreen-deciduous trees in mountains. On the other hand, *Humo-Ferric Podzols*, with well-differentiated horizons (ashy A_2 and black B_h horizons), are almost always *Podzols* of degradation *(secondary Podzols)* resulting from the substitution of the native forest by a moorland with Ericaceae (mainly heather), after clearing or gradual destruction of the forest by man (Guillet, 1972).

Local drainage conditions are important. If an acid water table develops, podzolization is accelerated by reduction of iron to the ferrous state (ferrous iron is much more mobile than ferric iron). Ground-water Podzols, with a "water table" resulting from rainfall (Stagnogleys) or recharged by ground water (sandy plains), display very particular morphological and geochemical features. They form the "Hydromorphic" subclass.

Depending upon the nature of the parent material and topographic position, there are two possible soil-forming processes that control the transition from *Brunified soils* (Chapter V) to *Podzolized* soils: (i) *direct* podzolization results from the direct action of acid organic matter on silicates; and (ii) *indirect* podzolization is preceded by a phase of clay and iron eluviation, together with a general acidification of the profile and a modification of the organic matter.

Podzolization under Atlantic climate (well-drained medium)

Direct podzolization on hard and acid rock. Under Atlantic climate, the most common "climatic" forest soils belong to the "Brunified" soil class (Chapter V) and display variances according to the nature of the parent material ("analogous" soils). However, forest soils still at equilibrium (site climaxes) can be classified with "Podzolized soils" only when formed on extremely acidic parent materials low in weatherable minerals, clay, and free iron. Depending upon the nature of the parent material and on topographic conditions, these soils are classified as either *Brown Podzolic soils* or *Podzolic soils* (with "humic" phases in humid mountains). Degradation and destruction of this equilibrium by man produce secondary Humo-Ferric Podzols on the most quartzous materials only (e.g., Vosgian sandstone; Guillet, 1972).

Indirect podzolization on loose sedimentary parent material (sand to sandy loam). On such materials, leaching of exchangeable bases (acidification) and eluviation of clay and iron constitute a preliminary step that promotes podzolization. This initial stage, characterized by the formation of *Eluviated Podzolic soils* or *Podzolic soils*, takes place under forest cover, either in "forest site climaxes," on very sandy parent material, or in soils that have begun to degrade under partially cleared forest. As with direct podzolization, Humo-Ferric Podzols, with well-differentiated black B_h and rust B_s horizons, are secondary Podzols resulting from an intense degradation process influenced by Ericaceae (heath, heather). But, contrary to direct podzolization, the "spodic" horizons are generally formed above an argillic B_t horizon resulting from the "illuviation" of clay and iron during the forested stage. This B_t horizon is missing only on slopes because of lateral eluviation (Profile XII_2).

Local discrepancies to this general scheme may occur depending on the nature of the parent material:

(i) In calcareous material, the B_t horizon is replaced by a β horizon as defined in Chapter I.

(ii) In some heterogeneous materials, where a pervious sandy layer rests on less permeable materials (such as some clays with flint), the B_t horizon is replaced by a hydromorphic B_g horizon which has the characteristics of a pseudogley. The spodic B_h or B_hB_s horizons are formed at the base of the eluviated and still drained A_2 horizon. This constitutes the Podzolic soil (or Podzol) with pseudogley (Profile XI_4).

(iii) When the surface sandy A_2 horizon is very low in iron, either because of its original composition or because of leaching during the forested stage, the "spodic" horizon is humic only and the ferruginous B_s horizon does not form. This soil is then called a Humic Podzol. In cases where the material is exceptionally low in iron, Humic Podzols are at equilibrium with the forest close to its climax. This is a very particular type of site climax (Profile XI_5).

Podzolization under hydromorphic conditions

The "rise of the water table" caused by a change or a decrease in drainage is at the origin of a particular type of podzolic genesis, which is less dependent on climate and may be observed in equatorial regions, under warm and humid climate.

Under Atlantic climate, more intense weathering of silicate minerals, massive mobilization of iron through reduction and complexing, and finally penetration of organic matter into all horizons contribute to the formation of *Hydromorphic Humic Podzols* (often with gley). Iron may migrate over a short distance and accumulate in a *placic horizon* (thin wavy iron-pan) as a result of a drastic change in redox potential. Iron may, on the contrary, migrate over a considerable distance as a complexed form and, through oxidation, accumulate in the ground water seepage zone where it forms large concretionary masses (Podzols with hard ferruginous alios).

Under equatorial climate, hydromorphism is more pronounced and leads to a greater degree of podzolization and weathering of clay minerals. Mobilization affects not only iron but also aluminum which is released by the weathering of kaolinite (hydromorphic degradation of ferralitic materials; Turenne, 1975, unpublished Doctorate Thesis, University of Nancy I, 180 pp.).

Table 8.   Genesis of Podzolized Soils

I. NO OR LITTLE HYDROMORPHISM

Climate

1. Boreal or alpine: Boreal or alpine Humo-Ferric Podzol

2. Atlantic:

    A. Hard and acid rock (slopes): "Direct" podzolization

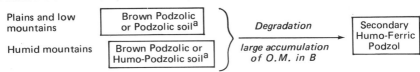

Plains and low mountains → Brown Podzolic or Podzolic soil[a]

Humid mountains → Brown Podzolic or Humo-Podzolic soil[a]

⟩ *Degradation* — *large accumulation of O.M. in B* → Secondary Humo-Ferric Podzol

    B. Loose sedimentary parent material (sandy loam): "Indirect" podzolization

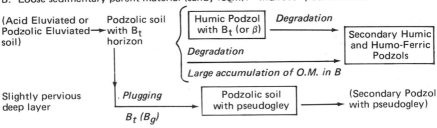

(Acid Eluviated or Podzolic Eluviated soil) → Podzolic soil with B_t horizon

Humic Podzol with B_t (or β) — *Degradation*

Degradation

Large accumulation of O.M. in B

→ Secondary Humic and Humo-Ferric Podzols

Slightly pervious deep layer — *Plugging* B_t (B_g) → Podzolic soil with pseudogley → (Secondary Podzol with pseudogley)

II. HYDROMORPHIC MEDIUM (acid ground water)

Atlantic or montane climate

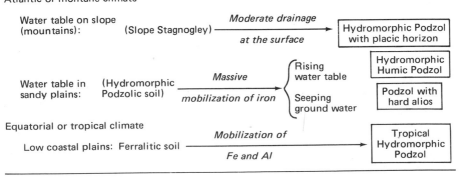

Water table on slope (mountains): (Slope Stagnogley) — *Moderate drainage at the surface* → Hydromorphic Podzol with placic horizon

Water table in sandy plains: (Hydromorphic Podzolic soil) — *Massive mobilization of iron*

⟨ Rising water table → Hydromorphic Humic Podzol

Seeping ground water → Podzol with hard alios

Equatorial or tropical climate

Low coastal plains: Ferralitic soil — *Mobilization of Fe and Al* → Tropical Hydromorphic Podzol

[a]Depending on the original iron content in the parent material and on topographic position.

XI₁: BROWN PODZOLIC SOIL
(Sol ocre podzolique)

F.A.O.: *Leptic Podzol;* U.S.: *Typic Haplorthod*
Location: Bruyères Communal Forest (Lot 10), Vosges, France.
Topography: Foot of slope, southwest exposure. Elevation 500 m.
Parent material: Triassic Vosgian sandstone, reworked by colluviation.
Climate: Low montane. P. 1,000 mm; M.T. 8.9°C.
Vegetation: "Rapaille,"* degraded oak and beech copse, replanted to Scotch pine. Acidophilous vegetation consisting of *Molinia coerulea, Vaccinium myrtillus, Calluna vulgaris.*

Profile Description

A_0 (4-0 cm):	Fibrous mor; many Ericaceae roots.
A_1 (0-10 cm):	Very dark gray (5 YR 3/1) loamy sand; single-grained; translucent quartz grains; clear boundary, emphasized by small ashy spots.
B_h (10-18 cm):	Dark reddish-gray (5 YR 4/2) loamy sand, loose; structureless; gravel; quartz grains coated with brown film.
B_s (18-45 cm):	Reddish-yellow (7.5 YR 6/6) loamy sand; friable "fluffy" structure, loose; gravel; many uniformly distributed roots; gradual boundary.
C:	Pink Vosgian sandstone, weathered to loose sand.

Geochemical and Biochemical Properties

Particle-size distribution: Mainly coarse sand with 12% silt and 6-7% clay. Clay minerals consist of equal amounts of kaolinite, inherited from the sandstone, and micaceous minerals. No evident clay eluviation (because of the slope).

Exchange complex: Very low content of exchangeable bases throughout the profile, but higher in A_0A_1 (2.8 meq/100 g) due to the biogeochemical cycle. S/T is very low throughout but slightly higher at the surface (4% in B_h and 8% in A_0A_1). pH (water) is 3.7 in A_0 and 4.5 in B_s.

Biochemical properties: Moder- to mor-type organic matter (22% O.M. in A_0 and 17% in A_1). Organic matter content is still 5% in B_h and 2.8% in B_s, but its nature is different: most of the organic components are extractable and fulvic acids predominate (1.5% FA). The C/N ratio is high, with values of 30 in A_1, 41 in B_h, and 20 in B_s, and constitutes a clear sign of podzolization.

Iron and aluminum: Moderate migration of iron which increases from 0.20% in A_1 to 0.55% in B_s. Migration of aluminum is more intense and extends to a greater depth (0.08% in A_1 and 0.53% in B_s). Under aerated conditions, aluminum complexes are more mobile than iron complexes. The aluminum/clay ratio reaches 0.08 in B_s, but this figure includes the translocated aluminum: the "estimated" ratio would be about 0.03 to 0.04. The complexed iron/free iron ratio reaches 0.27 in the B_h and B_s horizons which is characteristic of *spodic* horizons.

Genesis. This soil displays a moderate degree of podzolization. There is no clear A_2 horizon, the B_h horizon is still diffuse, and both are low in dark humic acids. Migration of iron is still moderate and restricted to the surface, while the B_s horizon remains loose and aerated. The "podzolic" evolution is related to the nature of the very quartzous parent material, which is low in free iron despite its color, and also to the beginning of a degradation by man (cutback to copses and pine plantation). However, podzolization has been slowed by two factors: (i) a biological factor—the presence of deciduous vegetation, (ii) a topographic factor— the slope which favors the lateral leaching of iron complexes, preventing the formation of an A_2 horizon. This type of soil is characteristic of the mixed evergreen-deciduous forest in low mountains, at the foot of slopes, on quartz sandstone low in iron.

REFERENCE: *Soil Map of France (1:100,000)*, Saint-Dié sheet, C.N.R.S., Nancy, 1973.

*Translators' note: *Rapaille* is a term used in the Vosges to designate the stand of a degraded forest on warm and dry exposures.

XI$_2$: PODZOLIC SOIL

(Sol podzolique)

F.A.O.: *Orthic Podzol;* U.S.: *Typic Haplorthod*

Location: Besange State Forest (Lot 52), Meurthe-et-Moselle, France.
Topography: Subhorizontal site on hilltop, 2% slope. Elevation 280 m.
Parent material: Rhaetian sandstone (Lower Lias).
Climate: Atlantic with continental trend. P. 720 mm; M.T. 10°C.
Vegetation: Acidophilous beech grove; *Pteridium aquilinum, Luzula albida.*

Profile Description

LA$_0$ (4-0 cm):	Incompletely decomposed litter.
A$_1$ (0-8 cm):	Gray to brown (7.5 YR 5/1) fine sand; fine crumb structure; fine roots; clear wavy boundary.
A$_2$ (8-25 cm):	Pinkish-gray (7.5 YR 6/2) fine sand; single-grained; horizon with variable thickness, wavy boundary.
B$_h$ (25-30 cm):	Brown (7.5 YR 5/2) fine sand; fine "coated" quartz grains, coprogenous aggregates; diffuse boundary.
B$_s$ (30-40 cm):	Yellowish-brown (10 YR 5/6) sand; gradual boundary.
C:	Fine sand, lighter than above; weathered sandstone.

Geochemical and Biochemical Properties

Particle-size distribution: Fine sand containing 8-9% silt. Intense translocation of clay (1.5% in A$_2$ and 6% in the B$_h$ and B$_s$ horizons). Migration of clay minerals and amorphous materials occurs simultaneously during the forested stage of podzolization (Guillet et al., 1975).

Exchange complex: The exchange capacity is very low except in A$_1$ (20 meq/100 g). S/T is very low in B (6%) but much higher in A$_1$ (28%). This indicates that the biogeochemical cycle remains efficient in the moder under deciduous vegetation. S amounts to 5.6 meq/100 g in A$_1$, but this horizon is extremely acid, as is the B horizon (pH in water is 3.5). This is related to the high production of soluble acid phenols in the slowly decaying litter.

Biochemistry: Low accumulation of organic matter in B (2%). The C/N ratio is never very high (16 in A$_1$ and 18 in the B$_h$ and B$_s$ horizons), in opposition to Podzols of degradation.

Iron and aluminum: The translocation index is typically lower for aluminum (1/15) than for iron (1/3.5) and reflects moderate podzolization. The free iron content in C is low (0.35%).

Genesis. This profile constitutes an intermediate stage of development between the Acid Brown soil (Profile IX$_1$) and the Humic Podzol under deciduous forest (Profile XI$_5$). It is located only 5 km from the Acid Brown soil. Litter composition (beech leaves) and parent material are identical, except that the free iron content is about twice as high in the sandstone of the Brown soil (0.7% versus 0.35%). Toutain (1974) has demonstrated experimentally that the iron content is responsible for the variations in the genesis of these two soils. In the Podzolic soil, phenolic compounds are not immediately insolubilized but migrate simultaneously with cations and clay minerals. This is "direct podzolization." It is moderate, because the organic matter originating from the deciduous vegetation remains *biodegradable* (low C/N ratio). The organic matter has a fast turnover and does not accumulate in the B horizon (Guillet, 1972).

REFERENCES: Guillet, 1972; Toutain, 1974.

XI_3: HUMIC PODZOLIC SOIL
(Sol podzolique humifère)

F.A.O.: *Orthic Podzol;* U.S.: *Typic Cryorthod*

Location: Gerardmer Forest, along Liaucourt road, Vosges, France.
Topography: Very steep slope (45%), west exposure. Elevation 950 m.
Parent material: Quartzous vein in primary granite.
Climate: Humid upper montane. P. 1,800 mm; M.T. 6.8°C.
Vegetation: Fir and spruce forest, with mor flora of *Vaccinium myrtillus, Deschampsia flexuosa, Dicranum scoparium.*

Profile Description

A_0 (5-0 cm):	Black, fine, slightly hydromorphic mor; sticky.
A_1 (0-10 cm):	Dark reddish-brown (5 YR 2/2) gravelly loamy sand; bare quartz grains and coprogenous aggregates; Ericaceae roots.
A_2 (10-37 cm):	Reddish-gray (5 YR 5/2) gravelly loamy sand; single-grained structure, loose, well aerated; common roots.
B_h (37-47 cm):	Dark reddish-brown (5 YR 3/4) sandy loam; weak blocky structure; soft to the touch.
B_{hs} (47-90 cm):	Heterogeneous horizon with blocky structure when undisturbed; breaks into a reddish-yellow (5 YR 6/8) mass with fluffy structure and yellowish-red (5 YR 4/6) diffuse elements, richer in organic matter, with coarser structure.
C:	Paler, loose gravelly sand.

Geochemical and Biochemical Properties

Particle-size distribution: Gravelly coarse sand with 12-14% silt. The profile shows evidence of clay formation with respect to the parent material. A moderate eluviation of clay (8% clay in A_2, 15% in B_h, and 7% in C) is partly offset by clay formation at the surface.

Exchange complex: Relatively high exchange capacity (minimum in A_2) due to a high humus content. The value of S is low in A_2, B_h, and B_{hs} (0.4 meq/100 g) but is much higher (6 meq/100 g) in the humic A_0 horizon. S/T varies less and is low throughout (8% in A_0 and 2% in B).

Biochemistry: Very high organic matter content throughout the profile (17% in A_1, 50% in B_h, and 7% in B_{hs}). The C/N ratio is relatively high (19 in A_1, 18 in B_h, and 28 in B_{hs}) but lower than in Podzols developed under Ericaceae. The organic matter accumulates *irregularly* in B, mainly as fulvic acids (3.8% fulvic acids in B_h; FA/HA ratio 4.5).

Iron and aluminum: This sand is relatively low in free iron (0.6%). The migration of iron is moderate (translocation index 1/4), but free iron content in A_2 is still relatively high (0.45%), so that this horizon does not present an "ashy" color. The translocation index for aluminum is lower (1/9) which is normal under aerated conditions.

Genesis. This soil is not a Podzol of degradation but a "climax" soil in equilibrium with the forest vegetation under montane climate on quartz sand low in iron. By opposition, on sand higher in iron, the climax soil is an Ochric Brown soil or even an Acid Brown soil (Souchier, 1971). The *slow biodegradation* of the organic matter is controlled simultaneously by the composition of the litter and the humid montane climate. All horizons are rich in organic matter. The abundant phenolic complexes migrate and accumulate in the B horizons, but are only slightly polymerized (predominance of fulvic acids). Although the profile is very humic, biochemical "podzolization" is still moderate. Thus, this is a "Podzolic soil," not a true Podzol: the A_2 horizon is not ashy because it contains free iron, the translocation index for iron is moderate.

REFERENCES: Souchier, 1971; Guillet, 1972.

XI₄: PODZOLIC SOIL WITH PSEUDOGLEY
(Sol podzolique à pseudogley)

F.A.O.: *Gleyic Podzol;* **U.S.:** *Aqualfic Haplorthod*

Location: Moulières State Forest (Lot 24), Vienne, France.
Topography: Platform with strong slope (10%). Elevation 135 m.
Parent material: Old alluvium with rounded gravel over truncated clay with flint.
Climate: Atlantic. P. 650 mm; M.T. 11.3°C.
Vegetation: Scotch pine planted on moorland with *Pteridium aquilinum*, *Calluna vulgaris*, *Erica cinerea*, *Ulex nanus*, resulting from the degradation of an oak forest.

Profile Description

A₀ (8-0 cm): Very dark brown (10 YR 2/2) fibrous mor; many roots; abrupt boundary.

A₁ (0-10 cm): Dark gray (10 YR 4/1) gravelly, humic sandy loam; single-grained structure with bare quartz grains and organic aggregates; many roots.

A₂ (10-30 cm): Gray (10 YR 5/1) gravelly sandy loam; single-grained structure.

B$_h$B$_s$ (30-40 cm): Dark yellowish-brown (10 YR 4/4) very gravelly sandy loam; massive and firm, cemented at places.

IIA$_{2g}$ (40-70 cm): Yellow (10 YR 7/6) loam to clay loam at depth; large yellowish-brown (10 YR 5/8) diffuse mottles; massive structure, slightly plastic.

IIB$_g$ (70-80 cm): Clay; brownish-yellow (10 YR 6/6) mottles and very pale brown (10 YR 8/3) vertical tongues; massive structure, plastic.

Geochemical and Biochemical Properties

Particle-size distribution: The profile clearly shows two layers: a gravelly sandy loam layer at the surface (4% clay and 30% silt in A₂) and a layer containing more clay but no gravel at depth (48% clay in IIB$_g$). The spodic horizon is slightly enriched with clay compared to the A₂ horizon (translocation index is 1/2).

Exchange complex: Base saturation is very low (10% in A₀, about 5% in the other horizons). The value of S is very low in A₂ and in B$_h$B$_s$ (0.30 meq/100 g), but much higher in A₀ (9.5 meq/100 g) due to the biogeochemical cycle. pH (water) is about 4 in the podzolic section of the profile, then 4.6 in the IIA$_{2g}$ and IIB$_g$ horizons.

Biochemistry: Mor with slow decomposition rate and a C/N ratio of 35. Organic matter is dispersed at depth throughout the podzolic section (2.6% in A₂ and 2% in B$_h$B$_s$), but drastically decreases in IIA$_{2g}$ and IIB$_g$.

Free iron: The iron content reflects the heterogeneity of the material. It is low in the podzolic section (0.2% in A₂ and 0.6% in B$_h$B$_s$ with a translocation index of 1/3); it increases greatly in IIA$_{2g}$ and exceeds 2% in IIB$_g$.

Genesis. This "podzolic" profile exhibits the early stages of an anthropic degradation under the influence of a secondary moorland containing heather, fern, and replanted Scotch pine. It has developed on a two-layer complex parent material. The surface layer has a coarse texture, relatively good drainage and is subjected to "secondary podzolization." The deeper layer has a fine texture and is subjected to hydromorphic conditions (decrease in porosity, plugging caused by clay illuviation—"indirect" type of podzolization). The profile is only "podzolic," since the A₂ horizon is not completely ashy (0.2% free iron) and the spodic horizon is weakly developed and discontinuous. This is a podzolic soil with hydromorphism at depth and is called a *Podzolic soil with pseudogley*.

REFERENCE: Unpublished work from Ecole Nationale des Eaux et Forêts, Research Station, 1958.

XI5: HUMIC PODZOL (with Beta horizon)

(Podzol humique)

F.A.O.: *Humic Podzol;* U.S.: *Alfic Haplorthod*

Location: La Tillaie, Fontainebleau Forest, Seine-et-Marne, France.
Topography: Plateau. Elevation 138 m.
Parent material: "Windblown" Stampian sand over Aquitanian limestone.
Climate: Moderate Atlantic (Parisian). P. 750 mm; M.T. 9°C.
Vegetation: Acidophilous beech grove; *Pteridium aquilinum, Lonicera periclymenum.*

Profile Description

A_0A_1 (0-4 cm):	Black fibrous mor; common roots.
A_1 (4-20 cm):	Very dark gray (10 YR 3/1) humic fine sand, becoming dark gray (10 YR 4/1) at depth; bare quartz grains; small coprogenous aggregates; clear wavy boundary.
A_2 (20-60 cm):	Light gray (10 YR 7/1) sand, with thin dark brown horizontal bands; single-grained structure; gradual wavy boundary.
B_h (60-75 cm):	Dark brown (7.5 YR 3/2) humic fine sand; light-colored spots, becoming gradually more ocherous; "coated" quartz grains; fine coprogenous aggregates; gradual boundary.
β (75-90 cm):	Brownish-yellow (10 YR 6/6) loamy sand; blocky structure; brown clay coatings.
IIC:	Loamy sand with limestone fragments.

Geochemical and Biochemical Properties

Particle-size distribution: Fine sand analogous to that of neighboring Profile X_2, but lower in silt, clay, and iron. This explains the higher degree of development of this profile. The β horizon contains 11.5% clay.

Exchange complex: Because exchange capacity is very low (except at the surface), base saturation is relatively high despite the low value of S (less than 1 meq/100 g, except in β which is base-saturated). S/T within the podzolic section of the profile ranges from 20% in A_1 to 28% in B_h (pH in water is 3.4 in A_1 and 4 in B_h).

Biochemistry: The organic matter decomposes slowly and accumulates in A_0A_1 and A_1: 30% O.M. in A_0A_1 (C/N ratio 32) and 9% in A_1 (C/N ratio 24). The migration of fulvic acids is still moderate and the organic matter accumulation does not exceed 2% in B_h.

Iron and aluminum: The β horizon has similar features, and the same origin, as the β horizon of Profile X_2. Mobile iron and aluminum contents are very low throughout the podzolic horizons. However, the translocation index for both elements (1/7) is relatively low (0.35% free iron in B_h).

Genesis. This soil is a true "Humic Podzol" and is therefore more developed than the Podzolic Eluviated soil (Profile X_2) or the Podzolic soil (Profile XI_2). The A_2 horizon is thicker and more ashy because the translocation index for iron is lower. As in the Podzolic Eluviated soil, *indirect* podzolization has been preceded by eluviation of iron and clay (see formation of the β horizon, Chapter I). Here, the spodic B_h horizon is almost in contact with the β horizon. The "eluviation" phase has "conditioned" the material for podzolization through the almost total removal of iron and clay. Again, the formation of this Podzol is influenced by parent material, rather than by vegetation. The organic matter turnover remains rapid in the B_h horizon.

REFERENCE: Robin, 1968. (Photo by A.-M. Robin.)

XI$_6$: ALPINE HUMO-FERRIC PODZOL
(Podzol humo-ferrugineux alpin)

F.A.O.: *Orthic Podzol;* U.S.: *Humic Cryorthod*

Location: Mount Corara, Carpathian Mountains, Romania.
Topography: Platform, with gentle slope. Elevation 2,100 m.
Parent material: Reworked periglacial material consisting of gneiss and various crystalline rocks.
Climate: Alpine. P. 1,250 mm; M.T. −1°C.
Vegetation: Moorland with *Loiseleuria procumbens*.

Profile Description

A$_0$ (5-0 cm):	Black (10 YR 2/1) fibrous humus.
A$_1$ (0-5 cm):	Very dark brown (10 YR 2/2), very humic; single-grained structure and coprogenous aggregates; translucid quartz grains; common roots; clear wavy boundary.
A$_2$ (5-12 cm):	Ashy, with grayish-brown (10 YR 5/2) spots; single-grained structure; skeletal quartz; rock fragments.
B$_h$ (12-18 cm):	Black (5 YR 2/1) humic accumulation; 50 μm aggregates cemented by a humo-ferric plasma alternating with coarse skeletal fragments; clear smooth boundary.
B$_{s1}$ (18-25 cm):	Dark brown (10 YR 3/3); same structure as above, less humic; gradual boundary.
B$_{s2}$ (25-45 cm):	Olive-brown matrix with dark yellowish-brown (10 YR 3/4) areas; gradual transition to yellowish-brown gravelly parent material.

Geochemical and Biochemical Properties

Particle-size distribution: Coarse texture; no evidence of clay translocation (10% clay in A$_2$ and 5% in B$_s$).

Exchange complex: S/T is low, especially in the lower horizons (10.8% in A$_0$ and A$_1$, but 2.5% in the illuvial horizons). pH (water) is relatively constant (4.3 to 4.5).

Biochemistry: Organic matter content is very high throughout the profile (35% in A$_1$, 5% in A$_2$, 26% in B$_h$, and 20% in B$_s$). FA/HA ratio is less than 1 in A$_1$, but increases with depth to reach 4 in B$_s$. C/N ratio is high throughout (38 in A$_0$, 24 in A$_1$, and still 27 in B$_h$ and 24 in B$_s$—a typical feature of podzolization).

Iron and aluminum hydroxides: Considerable migration of both elements: free iron content increases from 0.12% in the A horizons to 3.2% in B$_h$ (and 2.4% in B$_s$). Aluminum content increases from 0.1% in the A horizons to a maximum of 2.2% in B$_s$ (thus, aluminum migrates to a lower depth than iron).

Genesis. This soil is a *primary* Humo-Ferric Podzol in equilibrium with the alpine moorland. Although the morphological and biochemical properties of this soil are generally comparable to those of all Humo-Ferric Podzols with B$_h$ and B$_s$ horizons, some particular features oppose this profile to Atlantic Podzols of degradation. The profile is more compact and the ashy A$_2$ horizon is thin with respect to the spodic horizons. The accumulation of organic matter in the B horizons is greater than in Atlantic Podzols. Finally, the B$_h$ and B$_s$ horizons have less contrasting colors and structure than Atlantic Podzols under heath.

REFERENCE: *National Conference on Pedology: Mountain Soils*, Azuga Sinaia, Romania, 1969. (Photo by B. Souchier.)

PLATE XI

Podzolic Soils and Primary Podzols

XII₁: SECONDARY HUMO-FERRIC PODZOL
(Podzol humo-ferrugineux secondaire)

F.A.O.: *Orthic Podzol;* U.S.: *Typic Haplorthod*

Location: Saint-Dié-Taintrux, Vosges, France.
Topography: Steep slope (20%), southwest exposure. Elevation 580 m.
Parent material: Vosgian sandstone (Lower Trias), reworked at the surface.
Climate: Lower montane. P. 1,170 mm; M.T. 8.5°C.
Vegetation: Scotch pine; *Vaccinium myrtillus, Pteridium aquilinum, Calluna vulgaris, Stereodon schreberi.*

Profile Description

A_0 (10-0 cm): Dry fibrous mor with an F layer (6 cm) and a more humified H layer (4 cm); many roots.

A_1 (0-15 cm): Brown (7.5 YR 5/2) humic sand, some gravel; bare quartz grains in juxtaposition with organic aggregates; many roots; gradual boundary.

A_2 (15-35 cm): Light gray (5 YR 7/1) sand; single-grained structure, loose; some coarse sandstone fragments; clear boundary.

B_h (35-40 cm): Dark reddish-brown (5 YR 3/2) humic sandy loam; irregular horizon with *deep pockets*; quartz grains coated with black films; common roots at top.

B_s (40-50 cm): Reddish-yellow (7.5 YR 6/6) becoming lighter with depth; aliotic and cemented at top; quartz grains coated with ocherous films.

C: Light reddish-brown (5 YR 6/4) weathered Vosgian sandstone.

Geochemical and Biochemical Properties

Particle-size distribution: Coarse sand derived from weathered sandstone with 11-12% silt. Strong eluviation of clay with maximum clay content of 15% in B_h, versus only 2% in A_2. Only the micaceous minerals migrate; kaolinite, inherited from the sandstone, is coarser in texture and is not translocated. An isoquartz balance shows that a portion of the micaceous clay minerals in the A horizons has been destroyed and become "amorphous" (Guillet et al., 1975).

Exchange complex: Like in all Podzols, the exchange capacity is very high in A_0 (106 meq/100 g), high in B_h (29 meq/100 g), and very low in A_2 (4 meq/100 g). The value of S is less than 1 meq/100 g throughout the profile, except in A_0 and A_1 where there is a slight biological accumulation of cations. S/T is very low and ranges from 3 to 7%. pH values are 3.4 in A_0 and 4 in B_h and B_s.

Biochemistry: Organic matter is very abundant (10% in A_1, 2% in A_2, 8% in B_h, and 4% in B_s) and decomposes very slowly, as indicated by very high C/N ratios (43 in A_0, 30 in A_1, and 33 in B_h). Massive migration of mobile humus substances is indicated by a *FA/HA ratio of about 1 in B_h and between 4 and 5 in B_s.*

Iron and aluminum: Podzolization is favored by a relatively low free iron content (0.55%) in the parent material. Leaching of iron is important (translocation index 1/14) but not as strong as for aluminum (translocation index 1/80). The A_2 horizon still contains 0.13% free iron and 0.02% free aluminum.

Genesis. This Podzol of degradation (or secondary Podzol) results from the clearing of the evergreen forest and the subsequent takeover by heather. This took place some 2,000 years ago (Guillet, 1972). The original profile, which was in equilibrium with the forest, was probably a Brown Podzolic soil (or a Podzolic soil) similar to Profiles XI₁ and XI₂. During the forested phase, moderate podzolization and leaching of the micaceous clay minerals occurred simultaneously ("direct" podzolization). "Degradation" emphasized the migration of cations by formation of complexes and induced the destruction of a portion of the remaining clay minerals. Furthermore, the influence of heather on profile genesis is marked by the presence of a black B_h horizon which is rich in stable, polymerized humic acids.

REFERENCE: Guillet, 1972.

XII$_2$: SECONDARY HUMO-FERRIC PODZOL (with B$_t$B$_s$)
(Podzol humo-ferrugineux secondaire)
F.A.O.: *Orthic Podzol;* U.S.: *Alfic haplorthod*

Location: Versigny-lès-Usages, Aisne, France.
Topography: Top of sandy knoll, very gentle slope near a steep slope. Elevation 85 m.
Parent material: Thanetian sand, reworked at the surface.
Climate: Moderate Atlantic (Parisian). P. 680 mm; M.T. 9.7°C.
Vegetation: Acidophilous moorland with *Calluna vulgaris* (and scattered birch).

Profile Description

A$_0$ (6-0 cm):	Black (5 YR 2/1) fibrous mor at the surface, more humified with depth; many fine roots.
A$_1$ (0-15 cm):	Black (5 YR 2/1) to dark gray (5 YR 4/1) fine sand; bare quartz grains and juxtaposed organic aggregates; many roots, diffuse boundary.
A$_2$ (15-50 cm):	Light gray to gray (10 YR 6/1) fine sand; single-grained structure, loose; no roots (a strip of fragmented flint runs across this horizon); clear boundary.
B$_h$ (50-55 cm):	Black (10 YR 2/1) fine sand; massive structure, compact; coated quartz grains cemented by organic matter.
B$_t$B$_s$ (55-100 cm):	Fine sand; dark reddish-brown (5 YR 3/2) and dark brown to brown (7.5 YR 4/4) hard and cemented horizontal bands; upper bands richer in organic matter, lower bands richer in clay (plates).
C (100 cm +):	Light brown sand.

NOTE. The black B$_h$ and ashy A$_2$ horizons form pockets or even deep tongues in former root channels.

Geochemical and Biochemical Properties

Particle-size distribution: Fine sandy material with extremely low silt and clay contents (0.8%) throughout. However, the deepest reddish-brown bands contain 5-7% clay, which confers a mixed B$_t$B$_s$ character to this horizon.

Exchange complex: No available data. But Podzols usually have an exchange capacity that is high in A$_0$A$_1$, moderate in B, and almost nil in A$_2$. S/T is probably very low throughout the profile as indicated by a pH (water) less than 4 in all horizons. Most exchangeable bases are concentrated in A$_0$A$_1$ where they are strongly retained.

Biochemistry: A$_1$ contains 6% organic matter with a C/N ratio of 33. A$_2$ is almost devoid of organic matter (0.4%), while B$_h$ contains 4% organic matter with a very high C/N ratio (over 40). The horizontal aliotic bands have between 4 and 1% organic matter.

Free iron: The A$_2$ horizon is almost completely depleted of free iron (0.03%) and B$_h$ contains 0.5%. Iron content of "plates" ranges from 0.8 to 1%.

Genesis. This is a typical Podzol of degradation developed under a moorland with heather, at the site of a forest degraded by man. Heather is responsible for the formation of the "black B$_h$ horizon" which results from the migration and polymerization of stable polyphenolic complexes. Locally, these complexes have penetrated deeply into former root channels. The earlier *forested phase* was characterized by two processes: (i) clay and iron eluviation which normally generates a B$_t$ horizon—it is weakly expressed here due to the topographic position (lateral eluviation; see Profile XII$_3$); and (ii) moderate podzolization causing migration of small amounts of amorphous organomineral complexes and the remaining clay. This second process appears to be responsible for the formation of the bands (or "plates") in the upper part of the B$_t$B$_s$ horizon. This likely suggests indirect podzolization.

REFERENCE: *Soil Map of France (1:100,000)*, Laon sheet, Station agronomique de l'Aisne, 1973.

XII$_3$: HUMIC PODZOL WITH GLEY
(Podzol humique à gley)

F.A.O.: *Gleyic Podzol;* U.S.: *Alfic Haplaquod*

Location: Versigny-lès-Usages, Aisne, France.
Topography: Eocene cuesta piedmont, gentle slope, at edge of marshland. Elevation 75 m.
Parent material: Colluviated and reworked Thanetian sand.
Climate: Temperate, moderately Atlantic. P. 680 mm; M.T. 9.7°C.
Vegetation: Wet, acidophilous moorland with *Erica tetralix, Molinia coerulea.*

Profile Description

A$_0$A$_1$ (0-12 cm): Hydromoder; black organic layer (3 cm thick) overlying black (5 YR 2/1) sand with translucent quartz grains; common fine roots.

A$_2$ (12-18 cm): Pinkish-gray (5 YR 6/2) sand; single-grained structure, loose; diffuse boundary.

B$_h$ (18-35 cm): Dark brown (7.5 YR 3/2) sand, becoming lighter with depth; loose, porous; coated quartz grains and fine coprogenous aggregates; bounded by a very dark gray (5 YR 3/1) layer.

BG (35-50 cm): Sandy loam at top; blocky to massive structure; yellowish-red (5 YR 5/8) mottles increasing in size with depth, alternating with olive (5 Y 5/3) vertical streaks.

CG: Gradual change towards yellowish-brown "mottled" sand.

Geochemical and Biochemical Properties

Particle-size distribution: The entire upper part of the profile consists of *fine sand* with very low clay and silt contents (average 4%). The upper part of BG is enriched with clay (17%) as a result of *lateral eluviation probably preceding podzolization* (see Profile XII$_2$ which belongs to the same toposequence).

Exchange complex: Data not available. pH (water) increases with depth (3.8 in A$_0$A$_1$, 4.3 in B$_h$): pH is higher in the BC horizon (4.6), suggesting an increase in base saturation.

Biochemistry: Very high organic matter content at the surface (15% in A$_1$) but low in B$_h$ (3%). In A$_1$, the organic matter consists of fine aggregates in juxtaposition with quartz grains, whereas in B$_h$, it consists of fairly well-polymerized humic acids (dark color) coating the quartz grains. Note the change in the C/N ratios from a high value (30) in A$_0$ to a low value (12) in B$_h$.

Hydroxides: Free iron content is extremely low throughout the "podzolic" section of the profile (even in B$_h$, it is only 0.02%). It increases greatly in the BG and CG horizons to 0.75 and 0.62%, respectively. Aluminum content was not measured, but Jacquin et al. (1965) have reported an abundance of aluminous complexes in the B$_h$ horizons of similar soils in the Landes region.

Genesis. This profile is located at the foot of the sandy knoll where the preceding Podzol was described. The upper horizons are characteristic of Hydromorphic Podzols developed on sand with laterally moving acid ground water. *Iron is almost totally reduced and leached out laterally as complexes.* The A$_2$ is barely distinct from the A$_1$ horizon. The B$_h$ horizon has developed from the sole precipitation of strongly polymerized aluminous complexes, as the more mobile complexes are leached out laterally. The C/N ratio in the B$_h$ horizon is always lower than that in well-drained Podzols. This profile is characterized by a gley rich in iron (partly oxidized, partly reduced) and differs from Podzols in sandy plains where the gley is often "white" and low in iron. Here, the deep horizon is enriched with iron and clay through lateral input from the adjacent slope. Part of this iron is immobilized in the BG horizon due to its higher clay and exchangeable base contents.

REFERENCE: *Soil Map of France (1:100,000),* Laon sheet, Station agronomique de l'Aisne, 1973.

XII$_4$: PODZOL WITH CEMENTED FERRUGINOUS ALIOS
(Podzol à alios ferrugineux durci)

F.A.O.: *Ferric Podzol;* U.S.: *Typic Ferrod*

Location: Facture, Gascony Moorland, Gironde, France.
Topography: Gentle slope, ground water seepage zone, at edge of valley.
Parent material: Landes sand.
Climate: Aquitanian Atlantic. P. about 1,000 mm (relatively dry in summer);
M.T. 12.2°C.
Vegetation: Plantation of maritime pine on mesophytic moorland with *Ulex galii, Calluna vulgaris, Erica scoparia, Pteridium aquilinum.*

Profile Description

A$_0$ (6-0 cm):	Brownish-black fibrous mor; translucent quartz grains; many roots.
A$_1$ (0-6 cm):	Gray (7.5 YR 5/0) sand; single-grained structure; bare quartz grains and small organic aggregates; gradual boundary.
A$_2$ (6-20 cm):	Light gray (7.5 YR 7/0) sand; single-grained structure, loose; abrupt boundary.
B$_s$ (20-100 cm):	Yellowish-red (5 YR 5/8) very hard, cemented "alios"; presence of former root channels with general vertical orientation, may be empty or filled with gray to light gray sand as in A$_1$A$_2$; alios is dark reddish-brown (5 YR 3/3) around root channels.
C:	Pale yellow Landes sand.

Geochemical and Biochemical Properties

The texture of the parent material is dominated by very quartzous coarse sand and is characteristic of Landes sand. Silt plus clay content is very low (2 to 3%). The upper horizons are quite similar to those of Humo-Ferric Podzols found at "well-drained" and relatively dry sites. The mor is extremely acid (pH about 4), very low in exchangeable bases and has a C/N ratio over 30. The A$_2$ horizon has a very low exchange capacity and is almost depleted of iron (0.03%).

The B$_s$ horizon contrasts strongly with the upper horizons by its aspect, its structure, and its composition. It is less acid (pH in water 5.2), extremely rich in crystallized hydrated iron oxides (12.4% of goethite), and contains small amounts of organic matter (3-4%), mainly in the darker zones around root channels.

According to Schlichting (1965), the aluminum content of this type of cemented alios is much lower than that of a hydromorphic humic alios. Conversely, the accumulation of manganese is considerable.

Genesis. This soil has clearly developed in a so-called "ground water seepage zone" resulting from a sharp change in topography at the edge of a valley. In uphill positions, the acid water table fluctuates little and remains near the surface. Under such conditions and because ground water circulates slowly, iron is reduced and leached out laterally (see Profile XII$_3$, Hydromorphic Humic Podzol). However, in the profile under study, the water table declines so that oxidation can take place. Fe^{2+} and Mn^{2+} accumulate by precipitation in the oxidized and insoluble state. The complexing organic matter is transformed into fulvic and humic acids, but is gradually "diluted" by renewed supplies of iron. The organic matter content is then too low to prevent the crystallization of iron into goethite and the cementation of the alios (Schwertmann, 1966; Schwertmann et al., 1974).

REFERENCE: Unpublished report from the National Center for Forestry Research, Nancy, Bordeaux Field Station, 1964.

XII₅: HYDROMORPHIC PODZOL WITH PLACIC HORIZON
(Podzol hydromorphe à horizon placique)

F.A.O.: *Placic Podzol;* U.S.: *Spodic Placaquod*

Location: Grinden, Schliffkopf, Black Forest, West Germany.
Topography: Platform with moderate slope. Elevation 850 m.
Parent material: Sand reworked by solifluction, overlying Lower Triassic sandstone.
Climate: Humid montane with subalpine trend. P. 2,100 mm; M.T. 5°C.
Vegetation: Secondary moorland with *Molinia* sp. (former forest, cleared during the Middle Ages).

Profile Description

A_0 (12-0 cm): Black (N 1/) platy hydromoor with fine humus; many roots; gradual boundary.

A_2 (0-13 cm): Grayish-brown (2.5 Y 5/2) sand; single-grained structure; gradual boundary.

B_h (13-36 cm): Dark brown to brown (10 YR 4/3) humic sand; single-grained structure; common roots.

B_b (36-37 cm): Dark red (10 R 3/6) cemented "placic" horizon, forming an involute band.

B_s (37-63 cm): Red (2.5 YR 5/8) sand; quartz grains coated in brown; gradual boundary.

IIC: Pink Triassic sandstone in place.

NOTE. Profile XII₅ shows, in fact, a split profile: in the central portion, a weathered sandstone fragment is surrounded by a placic horizon; a *second placic horizon* can be found at the base of the profile.

Geochemical and Biochemical Properties

Particle-size distribution: Sand derived from periglacial reworked sandstone, with very little silt. Clay is absent in A_2 but represents 2-3% of the total soil in B.

Exchange complex: Exchange capacity is very low throughout the profile except in A_0. The exchange complex is so desaturated that the value of S is practically nil (traces). pH (water) is 3.5 in A and 4.2 in B_s.

Biochemistry: The A_0 horizon is totally organic with a C/N ratio of 38. The placic horizon has an organic matter content of 2% and a C/N ratio of 38. Yet, the accumulation of organic matter in B_h is small (1.2%).

Iron and aluminum: In this hydromorphic and reducing environment, iron is leached more rapidly from A_2 than aluminum (0.01% free iron and 0.08% aluminum). The placic horizon is enriched mainly with iron (7.5% versus 0.37% aluminum). Half of the iron is present in the crystalline state (goethite).

Genesis. This type of soil is closely associated with *Acid Peaty Stagnogleys* often found in small depressions. However, Podzols are usually located on higher ground which allows for lateral elimination of reduced iron. Iron is very mobile in such hydromorphic environments rich in complexing organic matter. The genesis of these two soils is influenced by the cold and humid montane climate (potential evapotranspiration is low throughout the year) and by the presence of a slope water table with slow circulation of the ground water.

The formation of the *placic horizon*—an involute, cemented, fine aliotic band—is not yet well understood. Its location seems to correspond to a zone of abrupt change in the E_h value. *Above* this horizon, a very reducing and stagnant water table with slow lateral circulation exists during most of the year (except for two months in summer).

REFERENCE: Schlichting, *Field trip Guide B, "Pseudogley and gley: Development and use of hydromorphic soils,"* Joint Meeting of Comm. V and VI, Int. Soc. Soil Sci., Stuttgart, 1971.

XII₆: TROPICAL HYDROMORPHIC PODZOL

(Podzol hydromorphe tropical)

F.A.O.: *Orthic Podzol;* U.S.: *Typic Tropaquod*

Location: Pariacabo Savanna, French Guiana.
Topography: Level; precoastal sand bars. Elevation: a few meters.
Parent material: Quaternary sandy loam marine deposit.
Climate: Humid equatorial. P. 2,200 mm (dry season for 3 months); M.T. 26.2°C.
Vegetation: Low savanna with *Rhynchospora barbata.*

Profile Description

A_1 (0-15 cm):   Gray (7.5 YR 5/0) slightly humic sand; single-grained structure; bare quartz grains; common roots; gradual boundary.

A_1A_2 (15-45 cm):   Pinkish-gray (7.5 YR 6/2) sand; single-grained to massive structure; bare quartz grains; few roots, gradual boundary.

A_{2g} (45-70 cm):   Very pale brown (10 YR 7/3) sand; ocherous mottles, vertical brown streaks; clear smooth boundary.

B_h (70-80 cm):   Dark yellowish-brown (10 YR 4/4) loamy sand; blocky structure; firm, hard but breaks under pressure; clear boundary.

B_s (80-100 cm):   Yellowish-brown (10 YR 5/8) sandy loam with dark brown to brown (10 YR 4/3) hardened spots; blocky structure; gradual boundary.

C_g: Ferralitic material with red (2.5 YR 5/8) and ocher mottles.

Geochemical and Biochemical Properties

Particle-size distribution: Parent material consists of a ferralitized Quaternary sandy loam marine deposit, containing essentially quartz and kaolinite, with some muscovite (19% clay). The clay-size fraction practically disappears in A_2 (1.5%) where whatever remains is "vermiculitized." Clay content is 10.5% in B_h and 18% in B_s.

Exchange complex: Exchange capacity is very low throughout the profile. From 0.3 meq/100 g in A_2, it increases to 3 meq/100 g in B_h. The content of exchangeable bases is also very low in the mineral horizons and ranges from 0.2 to 0.3 meq/100 g (pH in water is 5.1 in A_1 and 4.8 in B_h).

Biochemistry: The organic matter distribution is typical of Podzols with 1.5% only in A_1, traces in A_2, 2% in B_h, and 1.5% in B_s. Fulvic and humic acids are weakly polymerized and their contents increase during rainy seasons (Turenne, 1970). *Hydromorphism leads to an intensive mobilization of the organic matter in the A_1 horizon.*

Hydroxides: Free iron and aluminum oxides are almost entirely leached from the A horizons and accumulate in B_h and B_s (in B_h iron represents 0.5% and aluminum 1.2%). The translocation index is about 1/55 for iron and 1/120 for aluminum.

Genesis. According to Turenne (1970), the coastal alluvial plain of Guiana is characterized by an "evolution sequence" of soils from Podzols (at higher elevations) to Ferralitic soils (at lower elevations). Podzols result from the degradation of the ferralitic material caused by the presence and the fluctuation of a water table. During the rainy season, the water table rises to the soil surface and favors the formation of mobile, complexing humus substances. Clay minerals are subjected to acid hydrolysis with formation of mobile iron and aluminum complexes. Soluble silica is eliminated with ground water flow. When the water table falls, iron and aluminum complexes are leached vertically and laterally, and accumulate within the spodic horizons as amorphous or cryptocrystalline substances. Turenne (1970) has observed a correlation between the loss of clay from the A_2 horizon and the formation of amorphous alumina (or even poorly crystallized gibbsite) in the B_h and B_s horizons.

REFERENCES: Turenne, 1970; Turenne, J.F., 1975, unpublished Doctoral Thesis, University of Nancy I, 180 pp.

PLATE XII

SECONDARY PODZOLS AND HYDROMORPHIC PODZOLS

HYDROMORPHIC SOILS

Soils that are characterized by a prolonged deficit of oxygen in the profile, caused by temporary or permanent saturation of the pores with water, resulting in a partial reduction and mobilization of iron, along with a slowed decomposition of the organic matter.

In some cases, a more or less long persistence of pseudosoluble organomineral complexes is observed. This occurs mainly in an acid medium and may induce an acceleration of the biochemical "podzolization" process.

Several types of hydromorphism will be observed depending on the local soil conditions. Therefore, any ecological classification of Hydromorphic soils must be based on conditions of hydromorphism which generate different pedogenic processes.

Most typical Hydromorphic soils are characterized by a free "water table," which saturates all pores, coarse and fine. In the case of *Gley soils*, the water table is deep, permanent, and recharged by ground water. *Pseudogleys* are generated by perched water tables which are most often temporary and are caused by an impervious layer (often a plugged B_t horizon) which restricts the deep infiltration of rainwater. *Stagnogleys* will develop if the perched water tables persist for long periods of time under very humid montane climates with low potential evapotranspiration.

Usually, subsoil water moves laterally and very slowly. This is even the case in some Stagnogleys, in spite of the fact that their name means "soils with stagnant ground water." In regions with warm climate and contrasting seasons, temporary and very shallow water tables cause the eluviation of clay particles from the upper part of the profiles which display a very sharp horizonation *(Planosols)*.

Some soils exhibit hydromorphic trends, but their features are not related to the presence of a true water table. These are *Pelosols*, which are very clayey ("inherited" clay) and whose hydromorphism is essentially linked to capillary saturation phenomena. Under such conditions, water is very strongly held and moves very little, and redox processes lack intensity. In the French system of soil classification, these soils are, nevertheless, included with Hydromorphic soils.

NOTE. During the warm season, after heavy rains, true, mobile, and "discontinuous" water tables may form at the surface, in the humic horizons or in the desiccation cracks of the structural horizons of Pelosols.

Pseudogleys: temporary perched water tables

Pseudogleys are classified according to the following two criteria: (i) *the nature of the slightly permeable plugged horizon*, which can either be a former B_t horizon of an Eluviated soil (secondary Pseudogley), a slowly permeable geological stratum (primary Pseudogley) or a very dense Paleosol, "fragipan" (polycyclic or "complex" Pseudogley); (ii) *the degree of maturation*. The first signs of hydromorphism appear as ocherous or rusty mottles in a pale matrix, under weakly acid conditions and when the temporary water table is of short duration *(Incipient* or *Immature Pseudogleys)*. When the subsoil water becomes more acid, rises near the soil surface and persists for a longer time, mobilization of iron intensifies. A "hydromorphic" humus (hydromoder or hydromor) builds up. The upper horizon gradually loses its color, while concretions form at depth *(Mature Pseudogleys;* Becker, 1971). Eventually, in the most extreme cases, degradation of clay minerals occurs with release of aluminum under the action of corrosive soluble organic compounds *(Podzolic Pseudogleys)*.

Stagnogleys: lasting perched water tables

Stagnogleys are found in mountains with cold and humid climate. Nearly all are of the "primary" type and develop in "two-layered" parent materials located in poorly drained depressions. Sometimes, bedrock constitutes the "floor" of the water table while arenaceous weathering products form the "water-saturated" zone.

Stagnogleys differ from Pseudogleys by the persistence of ferrous iron in the profile (or else complete leaching of iron from the upper horizons) and by a frequent accumulation of organic matter *(anmoor* or *peat)*.

The Immature Stagnogley, formed in a still calcium-rich environment, remains colored with ferric iron, as reduction is only partial. But mobilization of iron is beginning as illustrated by rust spots or bands. Under very acid conditions, iron is completely reduced and gradually eliminated from the upper horizons, while silicate minerals are simultaneously weathered ("biochemical podzolization"). In extreme cases, peaty humus builds up. Iron precipitates as ferric iron in slightly better aerated zones, either as concretions or as a thin, involute "placic" horizon. This Podzolic Humic Stagnogley is close to the "Molkenpodzol" of Kubiena (1953) and its genesis resembles that of Podzols with a placic horizon.

Gley soils: permanent water tables

Gley soils form *very rapidly* under *very reducing* and often weakly acid soil conditions, under the influence of a permanent water table with more or less wide fluctuations. Thus, they cannot be classified according to their degree of maturation, but rather by the ground water regime and the amplitude of the water table fluctuations. Moreover, genesis is different whether occurring in well-mineralized (calcic or calcareous) or in acid media.

Well-mineralized medium. Ferrous iron is slightly mobile. It migrates only to a limited extent, generally in an upward direction.

Water table with wide fluctuations and an aerated surface horizon: *Low Humic Mineral Gley soil* (Alluvial Gley soil).

Water table with wide fluctuations but only a portion of the humus is aerated (hydromull): *Humic Oxidized Gley soil.* The water table is low enough during the dry season to allow the capillary rise of iron as $Fe(CO_3 H)_2$ and its precipitation in an oxidized G_o horizon with concretions. Underneath is the greenish G_r horizon, rich in reduced ferrous iron.

Water table with small fluctuations producing an anmoor-type humus: *Reduced Gley soil with anmoor.* The water table does not decline sufficiently during the summer to allow the formation of a G_o horizon. The profile is uniformly reduced.

Water table with small fluctuations: Calcic Peat soil.

Acid medium. In opposition to the well-mineralized medium, here iron is much more mobile, as it is not only reduced but also complexed into pseudosoluble forms. It is often leached out laterally by the ground water when in movement (white gley; *Gley-Podzol*). The formation of Gley-Podzols is analogous to that of Hydromorphic Humic Podzols. But the fluctuations of the water table are too small to allow a true B_h *horizon* to develop. Gley is present directly below the acid anmoor-like humic horizons. In the most extreme cases, acid peat develops to a great extent above an iron-depleted mineral horizon of the "white gley" type. Acid Peat soils evolve under the effect of drainage by illuviation of abundant hydrosoluble compounds (formation of a B_h horizon).

Pelosols

Pelosols are characterized by temporary hydromorphic conditions resulting from capillary saturation, associated with weak reduction phenomena. During some seasons, free water moves in the surface horizon and within cracks. These soils are very rich in clay inherited from the parent material and have affinities with several soil classes: (i) Immature soils, since the often ferriferous clay minerals are only slightly weathered (low free iron content); (ii) Vertisols, but vertic features are weakly expressed because of the lack of a dry season (Atlantic temperate climates); and (iii) Hydromorphic soils in which free iron is present in low amounts, partly in a reduced state, and reoxidizes slowly (especially in transitional forms).

The Immature Pelosol is often vertic, since vertic "structural" features can be acquired very rapidly. The structural (B) horizon is olive-gray. This color is due to the clay but not to the small amounts of ferrous iron.

Evolution proceeds either towards "brunification" through superficial aeration under the influence of a forest mull (Brown Pelosol) or towards the formation of locally mottled "Pseudogleys." Among these, the most typical derive from a local degradation of some ferriferous clay minerals which release free iron. Part of the iron becomes oxidized in the more aerated zones and mottles appear within the argillic horizon.*

*Another type of Pseudogley-Pelosol results from the formation of a very superficial and temporary water table due to a surface lithological discontinuity.

Planosols

These soils have formed under warm, very contrasting climates in a clay-rich substratum consisting of argillaceous sediments or in a pedogenic horizon which underwent a different evolution (e.g., ferruginous, vertic, sodic, etc.). In the second case, Planosols are polycyclic or polygenetic soils.

The *very shallow* ground water moves laterally, more or less rapidly and may disappear quickly, because of the seasonal contrasts. It causes a rapid *depletion* of clay and iron in the upper part of the profile. The bleached albic horizon which then develops has an abrupt boundary with the underlying argillic horizon.

Intense and abrupt phases of desiccation of the profiles allow Planosols to be distinguished from Pseudogleys.

At high elevation, in climatic zones with a lower mean temperature, depletion may be accompanied by a simultaneous mobilization and redistribution of the organic matter. The organic matter then forms black coatings on ped surfaces *(Planosols with humic B_t horizon)*.

Table 9. Evolution in Relation with Hydromorphism

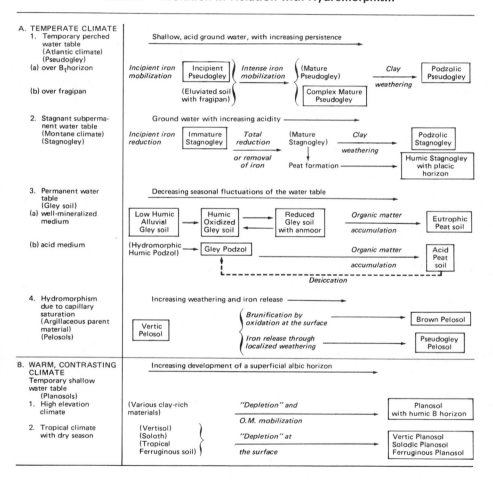

XIII₁: INCIPIENT PSEUDOGLEY

(Pseudogley initial)

F.A.O.: *Gleyic Luvisol;* U.S.: *Aquic Hapludalf*

Location: Dommartin-lès-Toul Forest, Meurthe-et-Moselle, France.
Topography: Level site, at the edge of a northwest slope. Elevation 280 m.
Parent material: Loam and old alluvium, over Bathonian marl.
Climate: Atlantic with continental trend. P. 737 mm; M.T. 9.4°C.
Vegetation: Oak forest (common oak) with hornbean; humid mull flora consists of *Poa chaixii, Deschampsia coespitosa.*

Profile Description

A_1 (0-5 cm): Dark brown to brown (10 YR 4/3) mesotrophic mull; fine crumb structure.

A_{2g} (5-20 cm): Very pale brown (10 YR 7/4) loam, with common mottles; very fine granular structure becoming massive with depth; few fine roots; gradual boundary.

B_g (20-45 cm): Yellow (10 YR 7/8) clay loam with much gravel; blocky to massive structure; bleached zones alternating with diffuse mottles; few roots; gradual boundary.

IIBC: Grayish-brown to light olive-brown (2.5 Y 5/3) clay, decarbonated marl, compact; less mottles with depth; gray, effervescent marl is found at 80 cm. (Note: This horizon is not shown in the photograph.)

Geochemical and Biochemical Properties

Particle-size distribution: The boundary between both materials is marked by the presence of gravel at a depth of 40-45 cm. Above this line, the texture of the alluvium is loam to clay loam (24-28% clay), while below, clay content exceeds 50%. Clay illuviation is clearly observed in the B_g horizon and the translocation index is approximately 1/1.4.

Exchange complex: Exchange capacity is fairly well related to clay content (12 meq/100 g in A_2, 28 meq/100 g in IIBC). Although very strongly acid (pH in water is 5 in A_{2g} and B_g), this soil is nevertheless much richer in exchangeable bases and less desaturated than the Acid Eluviated soil found upslope (Profile X_1). S/T is 62% in A_{2g} and 75% in B_g. This can be explained by the thinner gravelly loam alluvial layer and by the proximity of a calcium reserve in the underlying marl.

Biochemical properties: The mesotrophic mull in A_1 contains 6.5% organic matter. The C/N ratio of 15.5 is typical of a forest mull and decreases to 11 in the mineral B horizons, which is a feature of Brunified soils.

Iron and aluminum: The translocation index for iron is comparable to that for clay. An aluminum/clay ratio of 0.01 in A_2 indicates that clay minerals are less weathered than in the A_2 horizon of the Acid Eluviated soil.

Genesis. This is a *secondary Pseudogley* in which primary hydromorphism has been enhanced by a moderate translocation of clay, which has plugged the gravelly layer. But this is also an Incipient Pseudogley, because iron segregation is not yet very pronounced. Iron segregation shows up mainly as mottles within a pale brown matrix. It is likely that the relatively high base saturation slowed reduction and localized mobilization of iron. When compared to the neighboring profile (Hydromorphic Acid Eluviated soil, Profile X_1), this soil is more hydromorphic (mottles to the surface) but less acid (mesotrophic mull instead of an acid mull-moder). Consequently, clay minerals have not been weathered.

REFERENCE: Gury, M. *Legend, Soil Map of the Plateau of Haye,* C.N.R.S., Nancy, 1972.

XIII$_2$: COMPLEX MATURE PSEUDOGLEY WITH FRAGIPAN
(Pseudogley évolué complexe à fragipan)

F.A.O.: *Humic Gleysol;* U.S.: *Umbric Fragiaqualf*

Location: Charmes Forest, Vosges, France.
Topography: Plateau with very gentle slope (1.2%). Elevation 300 m.
Parent material: Loam and reworked, soliflucted material of a former terrace.
Climate: Atlantic with continental trend. P. 730 mm; M.T. 9.3°C.
Vegetation: Partially cleared forest with common oak; ground cover of *Molinia coerulea.*

Profile Description

A$_0$ (5-0 cm):	Black hydromor; fine structure; many fine roots.
A$_1$ (0-10 cm):	Black (10 YR 2/1) silt loam, becoming lighter with depth; blocky to massive structure; gradual boundary.
A$_{2g}$ (10-23 cm):	Gray to light gray (10 YR 6/1) silt loam, with darker humic spots; single-grained to massive structure; irregular black concretions.
B$_{tg}$ (23-37 cm):	Brownish-yellow (10 YR 6/6) silty clay loam, with white (10 YR 7/1) irregular pattern of bleached argillans; massive structure; many irregular concretions.
B$_x$ (37-150 cm):	Strong brown (7.5 YR 5/6) silt loam; very dense fragipan; vertical streaks filled with light gray (10 YR 7/1) argillans; massive structure.

Geochemical and Biochemical Properties

Particle-size distribution: The loamy material, which has been homogenized by cryoturbation (at least at the surface), contains about 15% sand. Clay content (15% in A) reaches a maximum in the B$_{tg}$ horizon (32%) and decreases to about 20% in the fragipan (translocation index 1/2).

Exchange complex: The exchange capacity is lowest in A$_{2g}$ (5 meq/100 g) and increases to 11-12 meq in A$_1$ and B$_{tg}$. S/T is low in A$_1$ (10%) and reaches 18% in B (pH in water is 4.5 in A$_0$, 4.7 in B).

Biochemistry: Slowly decaying organic matter, despite a medium C/N ratio of 20 in A$_0$ and A$_1$. This is due to the grass litter, relatively rich in N. The important penetration of organic matter in A$_1$ (10% O.M.) and in A$_2$ (2.5% with a C/N ratio of 20) is noteworthy.

Iron and aluminum: The iron content in A$_1$ (0.2%) and B$_{tg}$ (2%) demonstrates the very large mobilization of iron. The presence of concretions increases the average iron content in A$_2$ (0.8%). However, the free aluminum content, still low in B$_{tg}$ (0.35% or just over 1% on a clay basis), indicates that podzolic evolution remains weak.

Genesis. This is a complex soil with a polycyclic genesis. It is akin to Glossic Eluviated soils (Profile X$_3$) and has formed in a deep Paleosol, rich in iron (fragipan), underlying a layer homogenized by cryoturbation. This Pseudogley was most probably a "Hydromorphic Eluviated soil" under the natural forest cover. Deforestation generated a new process which involved the growth of *Molinia*, a reduction of the potential evapotranspiration rate, and a rise of the perched water table. The latter is very shallow (2-3 cm) from December to June. Under its influence, iron has been reduced, complexed with the hydrosoluble organic matter and leached out of the A$_1$ horizon. Iron has concentrated either in the concretions of the A$_{2g}$ horizon or in the B$_{tg}$ horizon, previously plugged by illuviated clay. Thus, this soil is a *secondary* and very *mature* Pseudogley with *a complex profile.*

REFERENCE: Becker, 1971. (Photo by M. Becker.)

XIII₃: PODZOLIC PSEUDOGLEY

(Pseudogley podzolique)

F.A.O.: *Gleyic Podzoluvisol;* U.S.: *Typic Glossaqualf*

Location: Laronxe, Mondon Forest, Meurthe-et-Moselle, France.
Topography: Level site. Elevation 260 m.
Parent material: Loam and reworked sand (old terrace).
Climate: Atlantic with continental trend. P. 750 mm; M.T. 9.4°C.
Vegetation: Largely cleared oak forest (common oak), *Rhamnus frangula* and ground cover of *Carex brizoides*.

Profile Description

A_0A_1 (0-25 cm): Gray to brown (7.5 YR 5/1) hydromoder with deep incorporation of organic matter consisting of aggregates between bare quartz grains; common roots.

A_{2g} (25-35 cm): Light gray (10 YR 7/1) sandy loam; black concretions; single-grained to massive structure.

B_g (35-75 cm): Reddish-yellow (7.5 YR 7/6) sandy loam; funnel-shaped bleached tongues with small black iron-manganese concretions along their edges; single-grained to massive structure.

B_{tg} (75-100 cm): Brownish-yellow (10 YR 6/6) sandy loam, more clay than above; thin vertical or diagonal tongues with bleached argillans.

C: Brown sand (Triassic sandstone from the terrace).

Geochemical and Biochemical Properties

Particle-size distribution: Homogeneous sandy loam material resulting from the reworking of the upper portion of the terrace. The silt fraction (30%) causes the plugging of the sand and enhances hydromorphism. Clay translocation is evidenced by a maximum clay content in B_t (15%, translocation index 1/2.5). This is a *secondary Pseudogley*.

Exchange complex: The average value of the exchange capacity is 10 meq/100 g. Base saturation is very low, around 7%. pH (water) is 4 in A_1, 4.3 in B_g.

Biochemical properties: The organic matter is deeply incorporated in A_1 (hydromoder feature) where its content is 5.3%. Some weakly polymerized components (FA) migrate into B_g (1.3% O.M.). The C/N ratio, which is 18 in the hydromoder, rises to 28 in B_g. These features are related to podzolization.

Iron and aluminum: Maximum iron (also Mn) content occurs in B_g, above the maximum in clay (B_t), which confirms the secondary character of the Pseudogley. Iron is eliminated from A_2 and concentrates in concretions and in B_g (3.1% Fe, translocation index 1/8). Aluminum is maximum in B_g, where the aluminum/clay ratio is 0.08. This is a sign of podzolization.

Genesis. This Pseudogley is highly developed under the influence of the hydromoder and the very acid ground water, which remains close to the surface throughout winter. In the A_{2g} horizon, iron is entirely mobilized by reduction and complexing with organic components. It accumulates as concretions (with Mn) above the clay-rich B_t horizon. This profile also displays some *signs of biochemical podzolization*. The presence of FA with a high C/N ratio, along with a high aluminum/clay ratio in the B_g horizon indicates biochemical weathering of the clay minerals. Very localized variations of redox potential in B_g should be noted. Iron and manganese are mobilized within the funnel-shaped "tongues" and precipitate along their edges. Such "functional tongues" are very different from inherited tongues which characterize the Glossic Eluviated soils formed on ancient material (Profile X₃).

REFERENCE: Souchier, B., C.N.R.S., 1972. (Photo by B. Souchier.)

XIII₄: COMPLEX IMMATURE STAGNOGLEY
(Stagnogley peu évolué complexe)

F.A.O.: *Gleyic Cambisol;* U.S.: *Aquic Eutrochrept*

Location: Raxplateau, Austrian Alps, Austria.
Topography: Edge of sinkhole, gently sloping. Elevation 1,840 m.
Parent material: Upper Jurassic white limestone, terra fusca, loamy colluvium.
Climate: Alpine elevation zone; "snowy vale." M.T. about 4°C.
Vegetation: Alpine meadow with hydrophilous plants.

Profile Description

A_0A_1 (0-8 cm):   Light yellowish-brown (2.5 Y 6/4) loamy hydromoder; very fine aggregates; many roots; clear boundary.
A_{2g} (8-24 cm):   Light olive-brown (2.5 Y 5/4) loam; ocher horizontal bands; massive to platy structure; firm and plastic, compact; clear boundary.
B_s (24-25 cm):   Reddish-yellow (7.5 YR 6/6) thin band of accumulation of ferric iron.
IIC (25-40 cm):   Dark brown to brown (7.5 YR 4/4) clay loam terra fusca; blocky to subangular blocky structure; plastic; white limestone gravel.

Geochemical and Biochemical Properties

Particle-size distribution: Texture reflects the heterogeneous nature of the material, which is further illustrated by variations in structure and porosity. Clay content in the compact loamy colluvium ranges from 13% (in A_1) to 22%. In the underlying terra fusca, clay amounts to 38%.

Exchange complex: The exchange capacity is relatively high throughout the profile (35 meq/100 g). The soil is moderately acid (pH in water 5.6). S/T which is 45% in A_1 and 30% in A_{2g}, approaches 100% in the IIC horizon, where traces of $CaCO_3$ are found in the fine earth fraction.

Biochemistry: The organic matter content is high in A_1 (17%) and in A_{2g} (4.6%) because of lack of aeration. However, the C/N ratio is below 15, as it is almost always in alpine meadows.

Iron: Iron is only partly reduced and therefore only partly mobilized. Free iron content increases gradually from A_1 to B_s. The translocation index varies from profile to profile from 1/3 to 1/4.

Genesis. Immature Stagnogleys are typical of "snowy vales" at the alpine elevation zone, formed on calcium-rich parent materials. In the present case, the parent material is complex and composed of two layers: (i) a colluviated loamy layer, highly compacted by percolating water at snowmelt, constitutes the "hydromorphic" section of the profile; (ii) at depth, a *terra fusca* layer which is more aerated and has a better structure constitutes a true Paleosol. The light olive-brown color of the A_{2g} horizon indicates that iron reduction is still incomplete under the moderately acid conditions. However, when reduced, iron migrates locally and accumulates not only at the surface of the plates, but mainly at the boundary with the terra fusca. There, it precipitates as ferric iron and forms a thin reddish yellow band. This shows a trend towards formation of a "Molkenpodzol" or Stagnogley with B horizon in the classification of Kubiena (1953) and Solar (1964).

REFERENCE: Solar, 1964.

XIII₅: PODZOLIC STAGNOGLEY
(Stagnogley podzolique)

F.A.O.: *Gleyic Podzol;* U.S.: *Typic Cryaquod*

Location: Ecorces Forest, Doubs, France.
Topography: Level site in the center of a small depression. Elevation 890 m.
Parent material: Colluviated fine loam over Sequanian marl.
Climate: Humid montane with subalpine trend. P. 1,850 mm; M.T. 4.5°C.
Vegetation: *Sphagno-picetum, Lycopodium annotinum, Polytrichum commune.*

Profile Description

A_0 (20-0 cm):	Very dark gray (10 YR 3/1) fibrous hydromor; many roots of all dimensions; clear boundary.
A_{2g} (0-20 cm):	Gray to light gray (10 YR 6/1) silt; single-grained to massive structure; few roots.
B_h (20-25 cm):	Slightly darker silt loam; trend towards blocky structure; precipitation of organic matter in the cracks.
B_g (25-55 cm):	Light brownish-gray (2.5 Y 6/2) clay loam with diffuse brownish-yellow (10 YR 6/8) mottles; plastic, compact.
IIG_r (55-90 cm):	Grayish-brown (2.5 Y 5/2) clayey gley, decarbonated marl; very plastic (not seen in photograph).
IIC:	Bluish-gray very calcareous Sequanian marl.

Geochemical and Biochemical Properties

Particle-size distribution: This is clearly a two-layer profile. The colluviated fine loam is very low in clay (2.5% in A_2) and quartz sand. It rests on a Sequanian marl which contains 58% of decarbonated clay down to 100 cm, but below that depth clay contains *62% $CaCO_3$.* A marked translocation of clay has occurred in the upper part of the profile (translocation index of 1/12 from A_2 to B_g).

Exchange complex: Exchange capacity is very high in A_0, almost nil in A_2, and reaches 12 meq/100 g in B_g. Low base saturation and *extreme acidity in A_0 and A_2* (S/T is 20% and pH is 4 in A_2). S/T increases progressively with depth to reach 45% in B_g and 100% below that horizon.

Biochemistry: Low biological activity. The hydromor contains 77% organic matter with a C/N ratio of 29. Slightly higher organic matter content (0.9%) in B_h with respect to A_{2g}.

Iron hydroxides: Iron migration is remarkable (0.03% iron in A_{2g} and 1% in B_g). Ferrous iron is present in B_g but especially in IIG_r. No data available for aluminum.

Genesis. This well-developed (even *Podzolic*) Stagnogley is typical of poorly drained forest sites at the subalpine elevation zone. It contrasts with the Ca-rich Immature Stagnogley of the alpine meadows. The spruce humus (very acid hydromor) provokes the complete mobilization of iron by reduction and complexing. It may even initiate podzolization of the quartz sand (formation of a weakly developed B_h, with a high C/N ratio). An opposition exists between the upper part of the profile and the underlying marl. The marl ensures an almost permanent water table but is not involved in the biogeochemical cycle of bases. Roots are restricted to the hydromor and nutrients are cycled between the vegetation and the humus.

REFERENCE: Richard, 1961.

XIII$_6$: HUMIC STAGNOGLEY (with placic horizon)

(Stagnogley humique)

F.A.O.: *Dystric Histosol;* U.S.: *Histic Placaquod*

Location: 130 km along the La Plata to Papayan road, Cauca, Colombia.

Topography: Gentle slope at the edge of a peat-bog depression. Elevation 2,850 m.

Parent material: Layered volcanic ash, deposited at different periods.

Climate: High elevation humid equatorial. P. 2,000 mm; M.T. 9°C.

Vegetation: Peaty zone with very few trees, in a depression surrounded by the high elevation forest; *Sphagnum.*

Profile Description

A$_0$ (14-0 cm): Very dusky red (2.5 YR 2/2) fibrous *Sphagnum* peat; permanently waterlogged.

A$_1$ (0-24 cm): Black (10 YR 2/1) clayey anmoor; massive structure, very plastic; waterlogged; abrupt boundary.

B$_b$ (25-26 cm): Dark reddish-brown (2.5 YR 3/4) placic horizon, very hard, somewhat involute; bounded on both sides by a reddish-yellow (5 YR 7/6) band with reddish spots.

IIC$_1$ (26-86 cm): Dark gray (10 YR 4/1) sandy loam with very diffuse red (2.5 YR 4/6) spots; massive structure becoming fine granular with depth; very thixotropic, very plastic.

IIB$_b$ (86-87 cm): Second placic horizon; similar to the upper one (not seen in the photograph).

Geochemical and Biochemical Properties

Particle-size distribution: The horizons overlying the B$_b$ horizon are essentially organic and the texture of the mineral fraction cannot be estimated. Below the B$_b$ horizon, the sandy loam contains 42% silt plus clay and 58% sand which derives mainly from the volcanic ash.

Exchange complex: It is highly desaturated. The exchange capacity is greater than 50 meq/100 g in A$_1$, while the value of S is only 1.7 meq/100 g. pH in water is extremely acid (about 4.1).

Biochemistry: The peaty horizon (71% O.M.) is underlain by an anmoor horizon with 28% organic matter (C/N ratio 20).

Iron and aluminum oxides: Iron and aluminum contents in A$_0$ and A$_1$, above the placic horizon, are very low (0.2% Fe, less than 0.1% Al). In the case of iron, this is due to mobilization through reduction. Most of the iron accumulates below the A horizons, in the placic horizon, where it crystallizes. The placic horizon also contains 6% organic matter.

Genesis. The two superposed profiles result from two evolution phases corresponding to volcanic ash deposits of different ages. The lower profile is a Hydromorphic Andic soil, while the upper one is a peaty-type Stagnogley with placic horizon. This "site climax" characterizes the poorly drained depressions of mountains with high precipitation and low potential evapotranspiration. It can be found at all latitudes. Reduced iron is mobilized as complexes and accumulates in the placic horizon at the boundary between the two materials. In the present case, the lower layer is composed of older and more deeply weathered volcanic ash than the upper layer. It contains more fine and amorphous elements and promotes the lateral circulation of the soil solution and the precipitation of well-crystallized iron.

REFERENCE: Faivre, P., C. Luna, and E. Ruiz, Codazzi Institute, Bogota, Colombia.

PLATE XIII

PSEUDOGLEYS AND STAGNOGLEYS

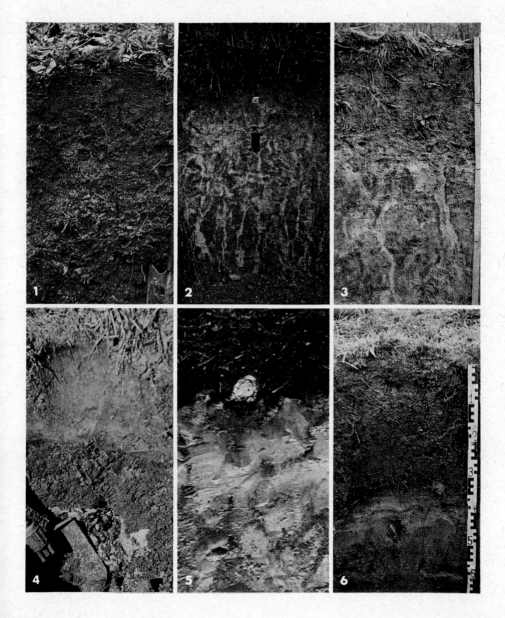

XIV₁: LOW HUMIC ALLUVIAL GLEY SOIL (AMPHIGLEY)
(Gley alluvial peu humifère)

F.A.O.: *Gleyic Fluvisol;* U.S.: *Agric Ochraquept*

Location: Flood-plain of the Rhine, Bietigheim, West Germany.
Topography: Rhine alluvial plain, level. Elevation 110 m.
Parent material: Recent Rhine alluvium.
Climate: Temperate. P. 730 mm; M.T. 9.5°C.
Vegetation: Humid meadow; alder clusters, poplar.

Profile Description

A_1 (0-16 cm): Very dark brown (7.5 YR 2/3) loam; crumb structure; common fine roots.

$(B)_g$ (16-40 cm): Light gray (10 YR 7/1) loam with small mottles and brown concretions; incipient blocky structure; clear boundary.

G_o (40-52 cm): Light gray (10 YR 7/1) sandy loam with large reddish-yellow (5 YR 6/8) mottles; gradual boundary.

G_{or} (52-80 cm): Light brownish-gray (10 YR 6/2) sandy loam with more diffuse and less distinct mottles; massive structure.

C: Grades into uniformly grayish alluvial sand. (Water table fluctuations range from the 30- to 150-cm depth).

Geochemical and Biochemical Properties

Particle-size distribution: Texture is moderately fine (20% clay, 40% sand) down to 40 cm where it becomes abruptly coarser (5% clay, 60% sand). Such abrupt textural changes, typical of alluvial deposits, reflect differences in the sedimentation rate.

Exchange complex: The exchange complex is base-saturated. No carbonates. pH (water) is 7.

Free iron: The maximum iron content (1.8%) occurs at a depth of 30-40 cm, then decreases progressively to 0.3% at 80 cm.

Organic matter: The organic matter content decreases rapidly with depth from 7% (C/N 10) in A_1, to 0.9% at 20 cm and to 0.4% at 50 cm.

Genesis. This Alluvial Gley soil is low in humus because of the wide fluctuations of the water table. Iron is incompletely reduced in G_{or}. Some of the reduced iron moves upwards in the profile and precipitates as large reddish-yellow mottles in the G_o horizon, in the water table fluctuation zone. Thus, the free iron content decreases from the subsurface to the deep horizons.

In fact, the hydromorphic evolution of this profile would be more complex, according to Schlichting and Blume (1971). The small dark brown concretions found in the B_g horizon above the G_o horizon indicate that, in this weakly structured loamy profile, the stagnation of rainwater causes temporary saturated conditions at the surface, unrelated to the formation of a true gley at depth. In the B_g horizon, partially reduced iron migrates over short distances in this poorly aerated medium and precipitates within concretions. This is a pseudogley-like horizon.

In summary, this soil could be more precisely designated as either "Pseudogley-Alluvial Gley soil" or "Alluvial Amphigley," as proposed by Romanian pedologists.

REFERENCE: Schlichting, E. and H.P. Blume, *Guidebook to field trip B, International Congress on Pseudogley and Gley*, Stuttgart, 1971.

XIV₂: HUMIC OXIDIZED GLEY SOIL
(Gley oxydé humifère)

F.A.O.: *Mollic Gleysol;* U.S.: *Typic Haplaquoll*

Location: Danube Plain, Platten, near Sentenhart, Germany.
Topography: Level bottom of a depression on a morainic plateau. Elevation 640 m.
Parent material: Colluvium of Riss moraine over Tertiary calcareous sandstone.
Climate: Lower montane. P. 850-900 mm; M.T. 7°C.
Vegetation: Humid meadow.

Profile Description

A_1 (0-20 cm):	Dark brown (7.5 YR 3/4) anmoor, becoming hydromull at the surface because of drainage; crumb to massive structure; plastic; common roots; precipitation of ferric iron in root channels.
G_o (20-40 cm):	Dark brown to brown (7.5 YR 4/2) sandy loam; oxidized gley with strong brown (7.5 YR 5/8) mottles; massive to prismatic structure; gradual boundary.
G_r (40-75 cm):	Gray to pinkish-gray (7.5 YR 6/1) sandy loam; reduced gley; massive structure; firm; grading at the base to gray (7.5 Y 5/1) weathered calcareous sandstone.
IICG:	Olive-gray calcareous sandstone.

Geochemical and Biochemical Properties

Particle-size distribution and clay minerals: Colluviated loam and fine sand resting on calcareous sandstone at 75 cm. This recent colluvium does not show any clay eluviation. Clay (mainly interstratified montmorillonite-vermiculite) is more abundant at the surface (10%) than in the G_o horizon (5%).

Exchange complex: The exchange capacity is relatively high in A_1 (24 meq/100 g) and decreases in G_o (16 meq/100 g). Base saturation is high throughout (92-95%) because of a large reserve of carbonates at depth and the capillary rise of Ca-rich water (pH in water is 6.6).

Biochemistry: The superficial hydromull is biologically very active whenever it is drained and aerated. It contains 11% organic matter with a C/N ratio of 12.

Iron hydroxides: Free iron contents show the upward migration of this element: it is highest in A_1 (2%) and G_o (2.7%) and lowest in G_r (0.35%). In the G_o horizon, iron forms coatings on the surface of peds (Blume, 1968).

Genesis. This soil is a partially oxidized Gley soil with hydromull, resulting from the recent and still incomplete transformation of a reduced Gley soil with anmoor through artificial drainage (see Profile XIV₃). With improved aeration, microbiological activity intensifies, an earthworm population develops and the production of CO_2 in the surface horizon increases considerably. Reduced iron, which is only slightly mobile in the presence of calcium, gives the G_r horizon an almost uniform bluish-gray or greenish-gray color (ferrous iron carbonates). Mobilization of iron is linked to the improved aeration of the humus. Carbon dioxide is brought in by rainwater and the soluble $Fe(CO_3H)_2$ which forms moves to the surface and, at the contact with oxygen, precipitates as reddish brown films of hydrated iron oxide on the surface of peds.

REFERENCE: Schlichting, E., *International Congress on Pseudogley and Gley. Guidebook to field trip B.* Stuttgart, 1971.

XIV$_3$: REDUCED GLEY SOIL WITH ANMOOR

(Gley réduit à anmoor)

F.A.O.: *Eutric Gleysol;* U.S.: *Typic Haplaquept*

Location: Ferme du Montet, Vandoeuvre-lès-Nancy, Meurthe-et-Moselle, France.
Topography: Bottom of depression, at foot of slope. Elevation 230 m.
Parent material: Silt loam colluvium, over Liassic marl.
Climate: Atlantic with continental trend. P. 730 mm; M.T. 9.5°C.
Vegetation: Humid meadow with rush and tall carex.

Profile Description

A$_1$ (0-10 cm):   Very dark grayish-brown (10 YR 3/2) anmoor; massive structure; sticky and plastic; many horizontal roots; clear boundary.
G$_r$ (10-30 cm):   Bluish-gray (10 BG 5/1) silt loam; massive structure; sticky and plastic; few roots; gradual boundary.
IICG$_r$ (30-70 cm):   Bluish-gray (10 BG 5/1) clay loam decarbonated marl; massive structure; sticky, very plastic; clear boundary (water table fluctuations range from 0 to 25 cm).

Geochemical and Biochemical Properties

Particle-size distribution: Silt loam colluvium (16% clay, 62% silt) over decarbonated Liassic marl (36% clay).

Exchange complex: The exchange capacity is very high in the anmoor (45 meq/100 g) and decreases to 33 meq/100 g in G$_r$. The medium is noncalcareous but almost base-saturated (pH in water 6.5), but local slight acidification and desaturation occur at a depth of 45 cm (pH in water 5.5).

Biochemistry: Calcic anmoor with 20% organic matter and a C/N ratio of 26. Good humification. Biological activity is slow, mainly because of a lack of aeration.

Iron hydroxides: The free iron content is 4% in IICG$_r$, but only 2% in A$_1$ and G$_r$, mainly because of a difference in the materials. Iron is not mobilized. Nguyen Kha and Duchaufour (1969) showed that ferrous iron predominates over ferric iron. In the slightly more acid zone (pH 5.5), 0.5% Fe^{2+} is exchangeable, which indicates the beginning of local mobilization under the influence of CO_2:

$$FeCO_3 + CO_2 + H_2O \rightarrow Fe(CO_3H)_2 \leftrightarrows Fe^{2+} \text{ (exchangeable)}.$$

Genesis. Because E$_h$ is seasonally very low (often negative), iron can become reduced even at relatively high pH values and in a Ca-rich medium. Iron is then in the ferrous state [$Fe(OH)_2$, $FeCO_3$, or other complex forms], but still insoluble and thus nonmobile. The production of CO_2 in the not very active anmoor is insufficient to mobilize iron and to induce its migration. Such a Gley soil is unstable and, when drained, changes rapidly, first into an Oxidized Gley soil with hydromull (formation of a G$_o$ horizon, see Profile XIV$_2$), then into a Brown soil with mull and a deep gley. The extreme rapidity of the oxidation processes is illustrated by ocher spots which form when a newly cut profile is left exposed.

REFERENCE: Nguyen Kha and Duchaufour, 1969.

XIV₄: HUMIC PODZOLIC GLEY SOIL

(Gley podzolique humifère)

F.A.O.: *Distric Histosol;* U.S.: *Histic Humaquept*

Location: Saint-Brisson peatland, Morvan Massif, Nièvre, France.
Topography: Gentle slope, east exposure, at edge of the peatland. Elevation 600 m.
Parent material: Granitic alluvium over granite.
Climate: Lower montane. P. 1,450 mm; M.T. about 8°C.
Vegetation: *Sphagnum, Molinia coerulea, Carex fusca.*

Profile Description

A_0 (20-0 cm):   Very dark grayish-brown (10 YR 3/2) peat changing to hydromor, organic matter partly decomposed into fine humus; usually waterlogged; plastic; common roots; gradual boundary.

A_1 (0-20 cm):   Dark grayish-brown (10 YR 4/2) very humic sandy loam; massive structure; plastic; usually waterlogged; clear boundary.

CG_r: White (10 YR 8/2) sand; massive structure; few roots (water table fluctuations range from 0 to 30 cm).

Geochemical and Biochemical Properties

Particle-size distribution: The humic horizons contain more silt and more clay (17%) than the mineral horizon, which is a sand.

Exchange complex: As the exchange capacity is related to the organic matter content, it decreases from 67 to 18 to 4 meq/100 g in the A_0, A_1, and CG_r horizons, respectively. Base saturation is 10% in A_1 and increases slightly to 17% in CG_r, but the value of S remains low (0.7 meq/100 g). pH (water) is 4.7.

Biochemistry: The surface horizon is an acid peat that has partly evolved into a hydromor (56% organic matter). This is evidenced by the almost complete disappearance of "fibers" and the fairly high content of mineral matter (C/N 19). The A_1 horizon still contains 14% organic matter with a C/N ratio of 14 and the CG_r horizon contains 8% O.M. with a C/N ratio of 12. Extractable carbon increases with depth from 18% in A_0, to 35% in A_1, and to 100% in CG_r. The FA/HA ratio increases with depth and reaches 2 in CG_r. The migration of slightly polymerized compounds (involving aluminum, see Profile XIV₆) is a sign of podzolization.

Iron hydroxides: The free iron content decreases regularly with depth from 0.90 to 0.12% in CG_r. Most of the iron is eliminated by reduction and lateral migration.

Genesis. Gley soils with acid anmoor (more or less peaty) contain less iron than calcium-rich Gley soils. Reduced iron is mobilized through complexing and eliminated from the gley by lateral migration. The gley becomes bleached (white gley). Capillary rise of iron is very limited and no G_o horizon can form. In addition, this type of profile often displays a podzolic character, as shown by the beginning of migration of soluble organic compounds. Aluminum forms complexes with these substances and migrates from the A_0 to the A_1 horizon (acting to some extent as a B_h) or even to the CG_r horizon. The same process can also be observed in some well-developed Acid Peat soils (Profile XIV₆).

REFERENCE: Menut, 1974. (Photo by G. Menut.)

XIV₅: WELL-DEVELOPED EUTROPHIC PEAT SOIL

(Tourbe eutrophe évoluée)

F.A.O.: *Eutric Histosol*; **U.S.:** *Typic Medihemist*

Location: Gillard Farm, Elba, New York, U.S.
Topography: Swampy depression connected to Lake Ontario. Elevation 220 m.
Parent material: Lacustrine calcareous alluvium (clay loam).
Climate: Continental trend. P. 685 mm; M.T. 9°C.
Vegetation: Vegetable crops.

Profile Description

A_{01} (0-40 cm): Black (10 YR 2/1) muck (or saprist); very fine granular structure; loose and friable; gradual boundary.

A_{02} (40-180 cm): Very dark gray (5 YR 3/1) partly decomposed hemist; fibrous organic debris (root, stem, leaf, and even wood fragments), embedded in a nonsticky and nonplastic fine humus.

G: Bluish-gray (10 BG 5/1) clay loam gley; massive structure; sticky and plastic (not seen on the photograph).

Geochemical and Biochemical Properties

This organic soil is free of contamination with external mineral particles. The ash content (15% in the surface horizon, 20% in the peat horizon) is in line with that generally found in eutrophic peat, which always contains more ash (of biological origin) than acid peat. Exchange capacity is higher in the *muck* (311 meq/100 g) than in the peat (240 meq/100 g). For a similar degree of evolution, eutrophic peat has an exchange capacity almost twice that of acid peat. Higher pH values and higher nitrogen and calcium contents favor the oxidation of lignin (*direct humification*) in the more aerated surface horizons.

At pH 6.2, base saturation (not given in the analysis) would be at least 50%.

The C/N ratio of calcic peat is always lower than that of undecomposed acid peat. Although not given, a C/N ratio of 20 in the peat and 14-15 in the muck may be estimated from analyses carried out in analogous soils of Europe.

Extractable carbon by pyrophosphate is 38% in the muck and only 9% in the peat.

Humification is enhanced by aeration. In particular, the *indirect humification* which results from the insolubilization of prehumus substances develops rapidly within the muck due to the high cation content. No migration is possible, and this explains the low extractable carbon values at depth.

Genesis. The transformation of calcic peat into "saprist" or "muck" under the influence of aeration at the surface is comparable to the transformation of acid peat into mor, under analogous conditions. But here, the processes are more complex, because of the more intense biological activity and the catalytic action of cations on the humification processes—on *direct humification* through oxidation of lignin (*inherited humin*) as well as on *indirect humification* through insolubilization of soluble prehumus substances (*fulvic acids, humic acids, and humin of insolubilization*).

REFERENCE: *Seventh International Congress on Soil Science, Guidebook, Vol. 1,* Madison, Wisconsin, 1960.

XIV$_6$: WELL-DEVELOPED ACID PEAT SOIL

(Tourbe acide évoluée)

F.A.O.: *Dystric Histosol;* U.S.: *Luvihemist*

Location: Saint-Brisson peatland, Morvan Massif, Nièvre, France.
Topography: Gentle slope, east exposure, at edge of depression. Elevation 600 m.
Parent material: Granitic alluvium over granite.
Climate: Lower montane. P. 1,450 mm; M.T. about 8°C.
Vegetation: Sparse stand of *Betula pubescens*, ground cover of *Molinia coerulea*.

Profile Description

A$_{01}$ (0-20 cm): Dark reddish-brown (5 YR 3/2) peat transformed into hydromoder due to the fall of the water table; fibrous to platy structure; many roots; clear boundary.

A$_{02}$ (20-40 cm): Dark reddish-brown (5 YR 2/2) fibrous peat, slightly decomposed and more humid; gradual boundary.

A$_0$B$_h$ (40-60 cm): Dark reddish-brown (5 YR 2/2) more decomposed peat, less fibrous (saprist); massive structure; very humid (water table at 60 cm), sticky; gradual boundary.

B$_h$ (60-70 cm): Black (5 YR 2/1) mixed organomineral horizon; massive structure; plastic.

G$_r$ (70-80 cm): Gray (5 YR 5/1) loam; massive structure.

Geochemical and Biochemical Properties

Particle-size distribution and mineral matter: The soil is almost exclusively organic to the 60-cm depth. But there is a slight contamination with quartz particles in A$_{01}$ (15% ash). A$_0$B$_h$ contains 7% ash. The organic matter content is only 20% in B$_h$, which has a clay loam texture.

Exchange complex: The exchange capacity is higher in the A$_0$ horizons (166 meq/100 g) than in the deeper organic horizons (140 meq/100 g). Base saturation is low (9% in A$_0$, 6% below) and is related to the extreme acidity (pH in water 4.2).

Biochemistry: Under the influence of vegetation, the C/N ratio is lower in A$_0$ (20), compared to the peat horizons, *sensu stricto* (34 in A$_0$B$_h$ and 30 in B$_h$). Extractable carbon content increases from 15% in A$_0$ to 22% in A$_0$B$_h$ and to 52% in B$_h$. FA/HA ratio also increases with depth.

Iron and aluminum hydroxides: Iron, which is moderately abundant in the peaty portion of the profile (0.25%), is almost eliminated from the G$_r$ horizon (0.07%). Aluminum content is 1% in A$_0$ and increases regularly with depth (2% in A$_0$B$_h$).

Genesis. This peatland has dried at the surface under the influence of woody plants and is undergoing a marked *podzolic evolution* close to that of some Acid Hydromorphic Humic Podzols. Genetic processes are comparable: (i) iron is eliminated laterally as soluble ferrous iron (white gley); (ii) prehumus substances, which form in A$_0$, form complexes with aluminum (of biological origin) and migrate to depth to form a true "spodic" B$_h$ horizon. This horizon is characterized by the simultaneous accumulation of amorphous free aluminum and FA+HA coming from the insolubilized prehumus substances. At the surface, *direct humification* proceeds through the oxidation of lignin in A$_0$ (increase of the exchange capacity) and, at depth, *indirect humification* occurs through the insolubilization of "prehumus substances" with high C/N ratio.

REFERENCE: Menut, 1974. (Photo by G. Menut.)

PLATE XIV
GLEYS SOILS AND PEAT SOILS

XV₁: VERTIC PELOSOL

(Pélosol vertique)

F.A.O.: *Vertic Cambisol;* U.S.: *Vertic Eutrochrept*

Location: Padoux Municipal Forest (Lot 15), Vosges, France.
Topography: Steep slope (25%), east exposure. Elevation 340 m.
Parent material: Lower Keuperian (Trias) red clay.
Climate: Atlantic with continental trend. P. 800 mm; M.T. 9.5°C.
Vegetation: Common oak forest, with beech and common maple. Active mull flora consists of *Ficaria ranunculoides, Arum italicum,* etc.

Profile Description

A_1 (0-18 cm): Very dark gray (5 YR 3/1) clayey mull; irregular crumb structure, becoming coarser at the base; compact; many roots; diffuse boundary.

(B)C (18-60 cm): Dark grayish-brown (10 YR 4/2) clay; prismatic structure breaking into coarse blocks; sticky, plastic, diagonal slickensides; wide desiccation cracks; few roots; clear boundary.

C: Very dark grayish-brown (10 YR 3/2) clay with 30-50% greenish patches; massive structure.

Geochemical and Biochemical Properties

Particle-size distribution: Very clayey (50% clay in A_1, 61% in C) with 10% sand. *Depletion of fine clay particles* in A_1 through lateral translocation (Nguyen Kha, 1973). Clay minerals, which are inherited from the parent material, are interstratified minerals involving a swelling clay. Traces of carbonates are found only in the C horizon.

Exchange complex: The exchange capacity is very high in A_1 (46 meq/100 g) and still high in (B)C (30 meq/100 g) because of the high content and the nature of the clay minerals. Base saturation is 100% despite a slightly acid soil reaction [pH is 6.2 in A_1, 6.6 in (B)C]. Magnesium content is high (6.4 meq/100 g in A_2, thus 4 times less than Ca^{2+}).

Biochemistry: Active mull with 12% organic matter with an abnormally high C/N ratio for a mull (19). The partly swelling clay minerals play an evident role in the evolution of the organic matter; the A_1 horizon is thicker and contains more humus than is usual for a mull. The content in *young* "humin" is especially high (Nguyen Kha, 1973).

Iron and aluminum hydroxides: Free iron content is 1.3% and free aluminum content is 0.6% in (B)C. Free iron corresponds to only 30% of total iron because some clay minerals (chlorites) are ferriferous.

Genesis. As this soil is on a slope, it is rejuvenated by erosion. Clay minerals are inherited, practically unweathered, from the C horizon, except for small losses of Fe and Mg through moderate hydrolysis of chlorites. *Vertic* features are definitely present (desiccation cracks, slickensides) but to a moderate extent, and they do not involve the organic matter (absence of a dry season). Similarly, *hydromorphic* features induced by short periods of capillary saturation are not very well expressed. The reddish color of the clay has disappeared, except in the C horizon and indicates an almost complete reduction of iron. The grayish-brown color of the whole profile is mainly due to the clay. The humus is a special mull of temperate climates, strongly influenced by the high clay content. It is very rich in young "humin" with a high C/N ratio.

REFERENCE: Nguyen Kha, 1973.

XV₂: BRUNIFIED PELOSOL

(Pélosol brunifié)

F.A.O.: *Gleyic Cambisol;* U.S.: *Aquic Eutrochrept*

Location: Parroy Forest, Meurthe-et-Moselle, France.
Topography: Plateau with gentle slope (5%), south exposure. Elevation 260 m.
Parent material: Keuperian (Upper Trias) variegated marl.
Climate: Atlantic with continental trend. P. 740 mm; M.T. 9.5°C.
Vegetation: Common oak, hornbeam, linden *(Tilia parvifolia)*. Ground cover of the humid mull consists of *Viburnum opulus, Deschampsia coespitosa, Ficaria ranunculoides.*

Profile Description

A_1 (0-5 cm):	Light gray mull; crumb structure; many roots; gradual boundary.
B_g (5-22 cm):	Grayish-brown (2.5 Y 5/2) clay loam; blocky to subangular blocky structure; small mottles; common roots; clear boundary.
IIB (22-45 cm):	Olive-gray (5 Y 5/2) clay; coarse blocky structure with shiny slickensides; few roots.
IIC:	Weak red (2.5 YR 4/2) variegated marl with horizontal gray bands; effervescent.

Geochemical and Biochemical Properties

Particle-size distribution: Two layers are present: the upper layer, or brunified section, consists of a 20-cm thick loam cover (37% clay in B_g); the deep layer is a clay derived from decarbonation of the marl (66% clay in IIB). The nature of the clay minerals is different in both materials: illite is dominant at the surface, while partly swelling interstratified minerals (montmorillonite-chlorite) are dominant in IIB.

Exchange complex: The exchange capacity reflects the composition of the parent materials (20 meq/100 g in B_g and 45 meq/100 g in IIB, a high value characteristic of the type of clays). The surface horizon with a pH value of 6 is slightly desaturated (S/T 90%). The lower horizons are base-saturated. Mg^{2+} is very abundant (22 meq/100 g in IIB, as much as Ca^{2+}).

Biochemistry: Very active mull with 5.5% organic matter and a C/N ratio of 16. This mull has a composition similar to that of the mesotrophic mull of temperate climates. It is much lower in clay-bound "humin" than the mull of the Vertic Pelosol.

Iron and aluminum hydroxides: In IIB, the amount of free iron and aluminum is low with respect to the high clay content. Free iron represents only 15% of total iron, and 0.7% on a total soil basis. This indicates a weak weathering of the magnesian and ferriferous clay minerals.

Genesis. Compared to the Vertic Pelosol, this soil is characterized by superficial brunification (with weak hydromorphic features as indicated by the occurrence of small mottles). Brunification is related to the presence, at the surface, of a soliflucted loam cover, more acid and moderately rich in clay. The Pelosol *sensu stricto* (the IIB and IIC horizons) has been "protected" by the loam cover and is less developed than in the preceding profile (Nguyen Kha, 1973). Because differences between wet and dry periods are less contrasted, structural features are less vertic. Clay minerals are barely weathered and the weathering index of free iron/total iron is lower than in the Vertic Pelosol.

REFERENCE: Bonneau et al., 1965.

XV₃: PSEUDOGLEY-PELOSOL

(Pseudogley-Pélosol)

F.A.O.: *Dystric Gleysol;* U.S.: *Vertic Haplaquept*

Location: Saint-Gobain State Forest, Aisne, France.
Topography: Level site, foot of moderate slope. Elevation 200 m.
Parent material: Reworked colluvium over Upper Lutetian clay.
Climate: Moderately Atlantic temperate. P. 680 mm; M.T. 9.7°C.
Vegetation: Common oak, aspen; hydrophilous flora with tall carex.

Profile Description

A_1 (0-5 cm):   Dark grayish-brown (10 YR 4/2) humid mull; crumb structure; many roots.

A_{2g} (5-22 cm):   Dark grayish-brown (2.5 Y 4/2) clay loam; medium blocky structure; small mottles; some gravel; clear smooth boundary.

IIB_g (22-80 cm):   Gray (5 Y 5/1) clay; medium blocky structure with yellowish-red (5 YR 4/6) coatings, few slickensides; gradual boundary.

IIC: Gray (5 Y 5/1) clay; prismatic to massive structure; fewer yellowish-red mottles than above.

Geochemical and Biochemical Properties

Particle-size distribution: It confirms the presence of two layers. Clay content varies from 30% in A_{2g} to 58% in IIB_g and IIC. Sand content does not exceed 5 to 6% in these horizons.

Exchange complex: The exchange capacity is moderate at the surface, but reaches 25 meq/100 g in the clayey horizons. Soil reaction is *extremely acid* (pH in water 4.5). Base saturation is approximately 40-45%, but is higher in the brunified surface horizon where pH reaches 5.2.

Biochemistry: The organic matter content is high (9.5% in A_1 and 4.6% in A_{2g}). This high content of active organic matter indicates a trend towards a hydromull.

Iron hydroxides: Free iron amounts to 1.5% in the clayey horizons. In the upper part of the profile, partial redistribution of free iron is due to leaching and hydromorphism (1% in A_{2g}, 2% in the upper part of IIB_g).

Genesis. This soil resembles Pelosols by its high content of gray clay-size particles (from 20 to 25 cm and below). As in the previous profile, the upper portion is brunified by the addition of allochthonous material. Furthermore, clay minerals have been weathered locally under the action of *acid* water flowing periodically from the surface along the desiccation cracks. Iron is then released and oxidized and forms coatings on the ped surfaces. This type of Pseudogley forms through alternate saturation and desiccation cycles of the clay mass. It is therefore different from Pseudogleys with a perched water table at the surface (Profiles XIII₁, XIII₂, and XIII₃).

REFERENCE: *Soil Map of France (1:100,000)*, Laon sheet, Laon Agricultural Station, 1973.

124   ECOLOGICAL ATLAS OF SOILS OF THE WORLD

XV₄: PLANOSOL (with humic Bₜ horizon)

(Planosol)

F.A.O.: *Eutric Planosol;* U.S.: *Typic Albaqualf*

Location: Sopo, Cundina Marca, Colombia.
Topography: Level, bottom lake deposits, "Bogota savanna." Elevation 2,600 m.
Parent material: Lacustrine clay deposits.
Climate: High elevation tropical, with dry season. P. about 1,000 mm; M.T. 13°C.
Vegetation: Currently, meadow with xerophilous trend (some Cactaceae).

Profile Description

A₁A₂ (0-25 cm):   Light brownish-gray (10 YR 6/2) loam, "albic" horizon; tendency to massive structure; slightly plastic; black concretions (1-3 mm); few roots, more common at the base; abrupt boundary.

Bₜ₁ (25-50 cm):   Light gray (10 YR 7/2) clay; strong prismatic structure with vertical cracks; *thick* black (7.5 YR 2/0) cutans on the ped surfaces; roots along ped surfaces; abrupt boundary.

Bₜ₂ (50-75 cm):   Light gray (10 YR 7/2) clay; coarse blocky with vertical cracks; well-expressed, dark brown to brown (7.5 YR 4/2) cutans, becoming strong brown (7.5 YR 5/8) at the base; clear wavy boundary.

C (75-100 cm):   Light gray (10 YR 7/1) lacustrine clay; massive structure, firm.

Geochemical and Biochemical Properties

Particle-size distribution: Abrupt textural change between the albic A₁A₂ horizon (about 20% clay) and the Bₜ₁ and Bₜ₂ horizons (54% clay). Clay content gradually increases towards the base of Bₜ₂ (72%), then drops to 42% in C. Clay minerals are poorly crystallized kaolinites with a small proportion of interstratified minerals. Clay minerals tend to disappear in A₂ due to their greater mobility.

Exchange complex: The exchange capacity is low in A₂ (9 meq/100 g) and moderate in Bₜ (23 meq/100 g). Except in A₂, the value of S is fairly high (16 to 19 meq/100 g in Bₜ with equal amounts of Ca²⁺ and Mg²⁺ and traces of Na⁺). S/T varies from 64 to 67%, but reaches 80% at the base of the profile (pH in water 5.3-5.6 depending on the horizon).

Biochemistry: The distribution of the organic matter is noteworthy. Its content is low (about 2%) and almost identical in A₁A₂ and in Bₜ₁. It is 1.2% in Bₜ₂. *The black cutans in Bₜ₁ contain 5% organic matter.*

Sesquioxides: The black cutans in Bₜ₁ contain little iron (0.4%) but more aluminum (0.9%). Manganese content is negligible. Ferrous iron is present in the core of the gray prisms (0.9%). As a whole, the Bₜ₂ horizon is enriched with iron (1.4%) and aluminum (0.8%).

Genesis. Like all Planosols, this profile is subjected to very sudden periods of surface saturation (rainstorms) contrasting with very dry periods. This induces a complex evolution of the profile which appears to be dominated by the anaerobic solubilization of the organic matter leading to dispersion of the fine clay particles on the one hand, and to complexing and mobilization of hydroxides on the other. Some of these compounds migrate downward to form the black and brown cutans of the Bₜ₁ and Bₜ₂ horizons. Another portion is eliminated laterally (depletion). The surface "albic" horizon contains almost exclusively fine quartz or residual silica. Its organic matter has been almost completely mobilized and eliminated vertically or laterally. Whatever the differences in general climate, the evolution of the organic matter, under seasonal anaerobiosis, is comparable to that of other soils: Hydromorphic Podzol (Profile XII₆) or even Gray Forest soil (Profile X₆).

REFERENCE: Faivre, P., Codazzi Institute, Bogota, Colombia, 1973. Mineralogical data by I.N.R.A., 1974. (Photo by P. Faivre.)

XV₅: SOLODIC PLANOSOL

(Planosol solodique)

F.A.O.: *Solodic Planosol;* U.S.: *Vertic Albaqualf*

Location: 23 km along the Neiva to Campoalegre road, Colombia.
Topography: Base of dejection cone, platform with gentle slope (2%). Elevation about 950 m.
Parent material: Middle Quaternary clayey alluvium-colluvium.
Climate: Tropical with a 3- to 4-month dry season. P. 1,620 mm; M.T. 24°C.
Vegetation: Grasses (meadow).

Profile Description

A_1A_2 (0-30 cm): Gray to light gray (10 YR 6/1) low humic silt loam; massive structure becoming coarse blocky at depth; friable; few roots; very fine black concretions.

A_2 (30-50 cm): White (10 YR 8/1) silty clay loam; firm in place, loose when broken; nonplastic; abrupt boundary.

B_t (50-80 cm): Black (10 YR 2/1) clay; columnar structure with black coatings, vertical cracks, top of columns are corroded; very firm, very hard when dry, very plastic when wet; gradual boundary.

C (80-250 cm): Dark brown (10 YR 3/3) clay loam.

Geochemical and Biochemical Properties

Particle-size distribution: The bleached silt loam (upper albic horizons), containing 17% clay and 53% silt, changes abruptly to a typical "solodic" horizon, with 51% clay. At the boundary, there is a deposit of fine silica particles of the size of clay (32%). The B_t horizon is characterized by a mixture of equal amounts of illite and montmorillonite.

Exchange complex: The exchange capacity is low in the A horizons, but increases greatly in the B_t horizon (22 meq/100 g) because of the high content of 2/1 clay minerals. Base saturation is 75% at the surface and increases to 90% in B_t and C. In B_t, Mg^{2+} amounts to 3.2 meq/100 g and Na^+ to 1.6 meq/100 g (or 8% of S). pH (water) is 6 in A_1A_2, 6.5 in A_2, and 7.5 in B_t.

Biochemistry: The organic matter content is remarkably low in the A horizons (about 1.5% with a C/N ratio of 9). The B_t horizon contains 1.7% organic matter, which is black, very polymerized and displays vertic and solodic features.

Iron oxides: Most of the free iron has been eliminated from the A_2 horizon (0.3%, in small concretions) but it is abundant in the vertic material (1.4%), as clay minerals are partly ferriferous.

Genesis. This is a typical Solodic Planosol where the albic horizon is enriched with residual silica and contrasts strongly with the underlying columnar solodic horizon. This soil results from the evolution of a Soloth, with a Na^+-rich "natric-like" B_t horizon, under the influence of a very superficial temporary water table. Ground water is responsible for the elimination of most of the Na^+ ions from the former natric horizon and has corroded the upper part of this horizon. It has also depleted the upper horizons of clay. The often unstable Na- and Mg-bearing clays have been either eliminated laterally (or even translocated vertically) as fine particles in suspension, or weathered. Residual silica has accumulated in the A_2 horizon, while mobilized iron has concentrated in small concretions. The B_t horizon displays *vertic* features related to the type of clay minerals and to the contrasts in soil climate.

REFERENCE: Faivre, P. and E. Ruiz, Codazzi Institute, Bogota, Colombia.

XV₆: FERRUGINOUS PLANOSOL (polycyclic)

(Planosol ferrugineux)

F.A.O.: *Eutric Planosol;* U.S.: *Typic Albaqualf*

Location: Caqueza, Giron de Blancas, Cundina Marca, Colombia.
Topography: Undulating platform, gentle slope (3%). Elevation 1,780 m.
Parent material: Clay alluvium, piedmont cone.
Climate: Tropical with dry trend, 5-month dry season. P. about 1,400 mm; M.T. 19°C.
Vegetation: Xerophilous grasses, Cactaceae.

Profile Description

A₁ (0-22 cm):   Dark brown (10 YR 3/3) slightly humic loam; weak crumb structure; few fine roots; diffuse boundary.

A₂ (22-50 cm):   White (10 YR 8/2) loam, albic horizon; medium blocky structure; slightly hard when dry; abrupt boundary, emphasized by a white band.

Bₜ (50-90 cm):   Yellowish-red (5 YR 5/8) and very dark grayish-brown (10 YR 3/2) clay; strong blocky structure, hard; peds are yellowish-red at the core and coated with dark grayish-brown films.

(B)C (90-230 cm):   Reddish-yellow (7.5 YR 7/8) clay, with dark brown spots; strong blocky structure, very hard when dry.

Geochemical and Biochemical Properties

Particle-size distribution: Texture is loamy in the surface horizons and suddenly becomes clayey (62% clay) in Bₜ. This indicates clay eluviation and strong depletion of the surface horizons (albic horizon). The Bₜ horizon resembles an argillic horizon subsequently enriched with cations and, possibly, with organic matter, as indicated by the structure of the coatings. Clay content decreases progressively in (B)C.

Weathering products: Typical of a Tropical Ferruginous soil, clay minerals consist mainly of kaolinite with lesser amounts of interstratified minerals and illite. Goethite and hematite contents are high. There is no free alumina.

Exchange complex: The moderate exchange capacity is related to the nature of clays (16-18 meq/100 g in Bₜ, or 26-29 meq/100 g of clay). Superficial planosolization has caused extreme acidification in A₁ and A₂ (S/T is 11-12%, pH in water 4.2), while the Bₜ and (B)C horizons are moderately acid [S/T is 66% in Bₜ, 83% in (B)C; pH 5.8].

Biochemistry: The organic matter content is low in A₁ (4%) with a C/N ratio of 11. The A₂ horizon still contains 1.5% organic matter and the Bₜ horizon 1.2%.

Genesis. This soil is clearly polycyclic. A slightly desaturated *Tropical Ferruginous soil* was subjected to superficial "planosolization" (depletion of all elements: clay, iron, exchangeable bases, etc.) under the influence of a very localized shallow water table. Adjacent profiles present a less planosolic aspect as they still contain clay and iron at the surface. Because of planosolization, the Bₜ horizon has apparently been subsequently enriched with iron and probably with organic matter, in the form of superposed (or incorporated) films on the preexisting clay coatings.

REFERENCE: Faivre, P., Codazzi Institute, Bogota, Colombia. Mineralogical analysis by I.N.R.A., Versailles, 1973.

PLATE XV

Pelosols and Planosols

CHAPTER VIII

FERSIALITIC SOILS

Soils developed under warm climate with a pronounced dry season, strongly colored by well-individualized iron oxides, with a saturated or only slightly desaturated exchange complex, and with 2/1 clays as the dominant clay minerals.

Although they are often grouped in the same class as Tropical Ferruginous soils, Fersialitic soils appear to possess their own identity as much because of their formation processes as by their ecology.

Fersialitic soils are characteristic of so-called "warm temperate" climates; in other words, climates with a lower mean annual temperature (ranging from 13 to 20°C) than that of Tropical Ferruginous soils and with very contrasted seasons. A considerable amount of rain falls during well-defined seasons which alternate with long dry spells. The *Mediterranean climate* is the most typical example, at least when it is sufficiently humid to allow a forest vegetation to grow (holm oak, cork-oak, Mediterranean oak with deciduous leaves). Drier zones with shrubby or herbaceous vegetation are characterized by other kinds of soil (*Reddish Chestnut* soils, Chapter IV). It must be emphasized that other regions of the world (such as California, Mexico, South Africa, and parts of Australia) have analogous climates which also promote fersialitization.

Fersialitization will proceed on any parent material provided it meets the following criteria: (i) absence of carbonates, or previous decarbonation, which is the general rule under forest cover; and (ii) sufficient supplies of calcium and magnesium other than carbonates, as well as enough weatherable minerals likely to release iron. Quartzous materials and acid granite are parent materials that cannot generate Fersialitic soils.

Genesis. The *genetic process* can be summarized as follows. As with Tropical Ferruginous soils, weathering is more intense than under temperate climate and similar parent materials release more iron. The free iron/total iron index usually exceeds 0.6, while it remains below this value in temperate soils. The long dry season promotes the conservation, or even the capillary "rise," of exchangeable bases in the profile. The medium does not become acid (or if it does, only slowly) and this favors the "conservation" in the soil profile of the products of weathering (silica, alumina, and iron). Clay minerals are thus dominated by 2/1 type clays and are either "inherited" (illite or preserved vermiculite) or result from neoformation (montmorillonite). Furthermore, two complementary soil-

127

forming processes take place: (i) rubefaction, whereby iron oxides are dehydrated during the dry season into "amorphous" or crystalline forms (hematite); and (ii) *eluviation* and especially *depletion* by selective erosion and lateral migration of fine clay particles. Only extremely young or reworked Fersialitic soils do not exhibit a B_t horizon.

The rate of fersialitization and of rubefaction depends on the ecological factors, mainly climatic conditions and type of parent material.

With respect to *climate*, the fastest rate occurs under the warmest Mediterranean climates (Greece, North Africa) and the formation of Fersialitic soils will tend to spread to parent materials of variable age and different nature. In the Mediterranean region of France, "rubefaction" is more gradual and Red soils are found only on old outcrops (*terra rossa* on hard limestone, old terraces).

Finally, under marginal climates, in particular, transitional Mediterranean climates (such as the Danubian climate of the Romanian plains or the Mediterranean-montane climate corresponding to the deciduous oak elevation zone), "rubefaction" is incomplete; in fact, it is only incipient. Under these conditions, the "climax" soil in equilibrium with vegetation and climate is a *more or less Eluviated Fersialitic Brown soil.*

Parent material also has an important influence on the rate of formation of Fersialitic soils. Calcareous sandstone high in silicate impurities is decarbonated and rubefied more rapidly than hard limestone which, because it weathers by "exfoliation," produces "terra rossa" (red clay resulting from decarbonation) at a very slow rate only (Lamouroux, 1971). Silicate rocks weather and become rubefied all the more rapidly as their iron and weatherable mineral contents increase.

NOTE. Two special types of genesis must be mentioned:

1. Some Fersialitic soils display a higher degree of development than normal in that *desaturation* of the profile is pronounced (acidification), and that depletion of clay (and even of silt and fine sand) in the upper horizons is more marked. On occasion, the proportion of kaolinite increases drastically because silica is lost from micaceous minerals or because 2/1 clay minerals migrate preferentially. Such soils can be found under relatively humid climate, on generally old parent material high in skeletal silica (Bottner, 1971).

2. The second phase relates to the opposite evolution, whereby an old and previously rubefied parent material (terra rossa) has been subsequently recarbonated through colluviation or mechanical reworking. This leads to Calcareous Brown soils on fersialitic material (this is the case of "sinkholes" in the Causses* region of France).

---

*Translators' note: *Causse* refers to calcareous table-land in Southern France.

Table 10.   Fersialitic Soils

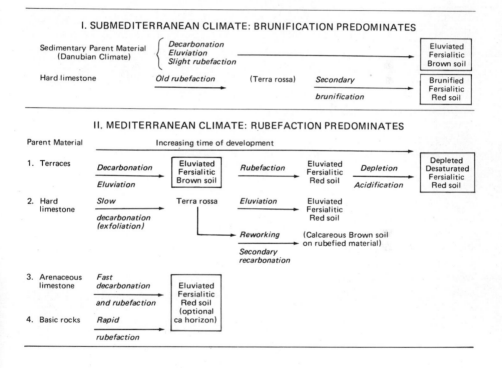

I. SUBMEDITERRANEAN CLIMATE: BRUNIFICATION PREDOMINATES

Sedimentary Parent Material
(Danubian Climate)
{ *Decarbonation*
Eluviation
Slight rubefaction }
→
Eluviated
Fersialitic
Brown soil

Hard limestone   *Old rubefaction* →   (Terra rossa)   *Secondary* → *brunification*   Brunified
Fersialitic
Red soil

II. MEDITERRANEAN CLIMATE: RUBEFACTION PREDOMINATES

Parent Material   Increasing time of development →

1. Terraces   *Decarbonation* / *Eluviation* →   Eluviated Fersialitic Brown soil   *Rubefaction* →   Eluviated Fersialitic Red soil   *Depletion* / *Acidification* →   Depleted Desaturated Fersialitic Red soil

2. Hard limestone   *Slow decarbonation (exfoliation)* →   Terra rossa   *Eluviation* →   Eluviated Fersialitic Red soil

→ *Reworking*   (Calcareous Brown soil on rubefied material)
Secondary recarbonation

3. Arenaceous limestone   *Fast decarbonation and rubefaction* →   Eluviated Fersialitic Red soil (optional ca horizon)

4. Basic rocks   *Rapid rubefaction* →

XVI₁: ELUVIATED FERSIALITIC BROWN SOIL (Danubian)

(Sol brun fersiallitique lessivé)

F.A.O.: *Chromic Luvisol;* U.S.: *Typic Paleustalf*

Location: Getic Piedmont, Podari, Romania.
Topography: Level plateau. Elevation 160 m.
Parent material: Pleistocene clay.
Climate: Submediterranean continental with weakly pronounced dry summer season. P. 550 mm; M.T. 10.8°C.
Vegetation: Under cultivation; formerly forested with *Quercus frainetto* and *Quercus cerris.*

Profile Description

A_p (0-18 cm): Grayish-brown (10 YR 5/2) clay loam; crumb to blocky structure; clear boundary.

A/B (18-52 cm): Brown (10 YR 5/3) clay loam; cubic to blocky structure; hard when dry, plastic when wet; gradual boundary.

B_{1t} (52-90 cm): Brown (7.5 YR 5/4) clay, reddish-brown (5 YR 4/3) when moist; strong cubic structure; hard when dry, firm when moist; thick clay skins; gradual boundary.

B_{2tg} (90-160 cm): Reddish-brown (5 YR 4/4) clay; cubic structure; hard; fine iron-manganese concretions; few bleached spots.

C_{ca}: Clay with powdery precipitated carbonates.

Geochemical and Biochemical Properties

Particle-size distribution: Clayey sediments containing 31% clay in A_p, 48% in B_{1t}; clay content varies little in the deeper horizons (45%). The translocation index is about 1/1.5. Depletion of clay by lateral migration appears to be greater than by vertical eluviation.

Exchange complex: The relatively high exchange capacity (27 meq/100 g at the surface and 34 meq/100 g in B_{1t}) results from the presence of micaceous minerals with maybe some montmorillonite of neoformation. Decarbonation is complete, but base saturation remains quite high [72-80%, pH (water) is 5.7] with mainly Ca^{2+} and Mg^{2+} on the exchange complex. NOTE: In other examples of this type of soil, S/T is close to 90%.

Biochemistry: Cultivation has homogenized the organic matter in the surface horizon. Organic matter content is 2.5% in A_p with a C/N ratio of 13, decreasing to 10 in the lower horizons. Iron content was not available.

Genesis. This type of soil was called "Reddish Brown Forest soil" by Romanian pedologists (Chirita et al., 1967; Mavrocordat, 1971). It is characteristic of the Danubian climate under forest, i.e., a subdued Mediterranean climate with a lower mean annual temperature and a weakly pronounced dry summer season. Under these conditions, the degree of development of the profile is intermediate between true fersialitization (rubefaction) and temperate brunification. Rubefaction is only incipient. Base saturation and pH are somewhat lower than in true Fersialitic soils. Nevertheless, the "depletion" of clay in the surface horizon, the relatively high exchange capacity of 2/1 clay minerals, and the reddish color of the B_t horizon are all attributes that differentiate this soil from temperate Brown soils, *sensu stricto.*

NOTE. Romanian pedologists consider some of these soils to be polycyclic soils with some features partly inherited.

REFERENCE: Oancea, C. and C. Rapaport, *Guidebook to field trip II, 8th International Congress of Soil Science*, Bucharest, 1964.

XVI₂: BRUNIFIED FERSIALITIC RED SOIL (terra rossa)

(Sol rouge fersiallitique brunifié)

F.A.O.: *Chromic Luvisol;* U.S.: *Typic Rhodoxeralf*

Location: "El Zegri" mountain pass, Bailen to Granada road, Spain.
Topography: Platform at top of hill. Elevation 1,000 m.
Parent material: Upper Lias hard limestone, "terra rossa."
Climate: Mediterranean with montane trend. P. about 600 mm; M.T. about 13°C.
Vegetation: Degraded forest of *Quercus lusitanica, Quercus ilex, Juniperus oxycedrus, Genista scorpius.*

Profile Description

A₁ (0-5 cm):   Reddish-brown (5 YR 4/4) mull; well-aerated crumb structure; strong biological activity; common roots.

(B) (5-25 cm):   Yellowish-red (5 YR 4/6) clay; well-aerated crumb structure; common roots; gradual boundary.

B_t (25-50 cm):   Red (2.5 YR 4/6) clay; coarse blocky to prismatic structure, shiny clay skins; very hard when dry; some weathered calcareous gravel.

C: Fractured Liassic limestone; pockets filled with terra rossa.

Geochemical and Biochemical Properties

Particle-size distribution and clay minerals: In the lower half of the profile, texture and clay minerals are typical of the red clay remaining after decarbonation *(terra rossa)*. The B_t horizon contains 85% clay, consisting mainly of illite and montmorillonite, and very little sand. However, the sand and silt fractions increase considerably towards the surface, where they represent approximately 50% of the fine earth. This is probably the result of *vertical eluviation* accompanied by extensive *lateral depletion*, but it is impossible to estimate the relative contribution of either process.

Exchange complex: The exchange capacity listed in the analytical data for this profile appears to be too low for a terra rossa. Normally, an exchange capacity of about 50 meq/100 g of clay can be expected. The exchange complex is almost saturated with Ca^{2+} and Mg^{2+} [S/T 90%, pH (water) is 6.9 in (B)].

Biochemistry: Due to the forest cover and the montane climatic trend, organic matter content is high in the upper soil horizons [14% in A₁ and 6% in (B)]. C/N ratio is 14 in A₁ and decreases rapidly with depth. Very intense biological activity.

Iron: High free iron content (3.2% in A₁ and 4.5% in B_t) which is characteristic of terra rossa. The free iron/total iron ratio is high (0.71) and is a sign of fersialitization. Free aluminum content is negligible.

Genesis. Under this type of climate, the terra rossa, resulting from slow exfoliation by dissolution of the hard limestone (Lamouroux, 1971), must be considered as a Paleosol serving as parent material. The recent genesis of the profile is influenced by the forest vegetation and the high elevation climate. Two processes can be distinguished: (i) pronounced brunification by incorporation of large amounts of organic matter and partial rehydration of iron oxides; and (ii) depletion at the surface of fine particles by both lateral migration and vertical eluviation. The B_t horizon exhibits an undeniable "argillic" character. Thus, this soil is a Fersialitic Paleosol which is brunified, depleted, and eluviated at the same time.

REFERENCE: Albareda, J.M., *Conference on Mediterranean Soils*, Center of Edaphology, Madrid, 1966.

XVI₃: ELUVIATED FERSIALITIC BROWN SOIL (recent terrace)

(Sol brun fersiallitique lessivé)

F.A.O.: *Calcic Luvisol;* U.S.: *Lithic Haploxeralf*

Location: Paravaudes-Beaucaire quarry, Gard, France.
Topography: Level site at foot of Costières hills. Elevation, a few meters.
Parent material: Recent Wurmian terrace; calcareous gravel.
Climate: Subhumid Mediterranean with dry season. P. 600 mm; M.T. 14°C.
Vegetation: Fallow land.

Profile Description

A_p (0-40 cm):   Light reddish-brown (5 YR 6/3) sandy loam; massive to blocky structure; siliceous gravel, more abundant with depth, slightly effervescent at places; common roots; clear boundary.

B_t (40-80 cm):   Reddish-brown (5 YR 4/4) sandy loam; blocky structure; very porous, firm; much rounded siliceous gravel, clay skins on gravel; abrupt boundary.

C_{ca}:   Calcareous crust, cemented gravel; massive structure, firm; friable calcareous cement; effervescent throughout.

Geochemical and Biochemical Properties

Particle-size distribution and carbonates: The amount of fine earth decreases with depth. This indicates the young age of the profile. Slight depletion probably compensated by deposition of silt at the surface. The C_{ca} horizon contains 26% $CaCO_3$. But decarbonation is almost complete in A_p and B_t (traces of $CaCO_3$) and has enabled clay to be *moderately* eluviated (10% clay in A_p and 18% in B_t).

Exchange complex: Relatively high exchange capacity (60 meq/100 g of *clay*) indicating the presence of 2/1 clay minerals. The exchange complex appears to be "supersaturated" (the analysis yields S/T over 100%). This is frequent in many Fersialitic soils high in calcareous reserves (Bottner, 1971) and could be due to the capillary "rise" of carbonates during the dry season. pH (water) is 8.5.

Genesis. This soil is a typical example of an immature Fersialitic soil developed on a *recent* alluvial terrace containing calcareous elements. It contrasts with more mature profiles found on *old* terraces even though parent material and climate may be similar. Rubefaction is incomplete and not very pronounced; there is no depletion of fine particles; decarbonation takes place only in the upper portion of the profile but is incomplete, so that if the analysis is performed carelessly the soil appears to be "supersaturated" with calcium (high pH values). Thus, a moderate vertical eluviation of clay particles is not prevented; in fact, eluviation begins soon after decarbonation under these climates. This profile shows that, in the South of France, rubefaction is slow inasmuch as it can be considered to be a modern process.

REFERENCE: *Soil Map of France (1:100,000),* Nîmes sheet, Compagnie du bas Rhône-Languedoc, Nîmes, 1974.

XVI₄: DEPLETED, DESATURATED FERSIALITIC RED SOIL
(old terrace)

(Sol rouge fersiallitique désaturé, appauvri)

F.A.O.: *Ferric Acrisol;* U.S.: *Ultic Palexeralf*

Location: Doscare quarry, Saint-Aunés, Hérault, France.
Topography: Level, at edge of terrace. Elevation 54 m.
Parent material: Early Quaternary terrace: quartzite and limestone rounded gravel, decarbonated to a depth of 9 m.
Climate: Subhumid Mediterranean with dry season. P. 700 mm; M.T. 14°C.
Vegetation: Holm oak *(Quercus ilex)* forest, partly cleared, formerly cultivated.

Profile Description

A_0A_1 (0-4 cm): Very dark gray forest xeromoder, fibrous.
A_2 (4-70 cm): (Former A_p to 44 cm). Pink (5 YR 7/4) very gravelly loamy sand; single-grained; very porous; 91% rounded unsorted gravel (quartzite and porous, decarbonated, and weathered limestone); common roots; *clear* smooth boundary.
B_{tg} (70-150 cm): Red (2.5 YR 5/8) gravelly clay with both dark and partially bleached mottles; blocky structure with red clay skins on peds; 69% gravel; gradual boundary.
B_g (150-200 cm): Variegated red (7.5 R 4/6) and white (10 BG 8/1) gravelly clay loam; blocky structure (Not visible in photograph).

Geochemical and Biochemical Properties

Particle-size distribution: The A_2 horizon is markedly "depleted" of fine earth (9% in A_2 versus 31% in B_{tg}) and especially of clay (5% on a fine earth basis in A_2 versus 47% in B $_{tg}$). On a total soil basis, the translocation index for clay would be 1/32, but this figure is meaningless in this instance. The still high clay content in B_g and below (35%) indicates that lateral depletion surpasses vertical eluviation.

Exchange complex: Exchange capacity ranges between 20 and 24 meq/100 g in B_{tg} and B_g. The surface horizon (A_2) is almost base-saturated as is the upper part of B_{tg}. This feature has been inherited from the former cultivation. However, below the 100-cm depth, the *value of S/T is 49% and pH (water) is 5.* Both these values are low for a Fersialitic soil.

Biochemistry: The forest xeromoder was not analyzed. However, data from other, but similar, soils suggest that the C/N ratio would be approximately 25. It decreases to 10 in the former A_p horizon (top of A_2) which contains 0.8% organic matter.

Genesis. This soil is an excellent example of a highly developed, desaturated Fersialitic soil having formed over a long period of time on very old and very pervious parent material. *Depletion* (i.e., lateral elimination by selective erosion of not only clay but of most of the fine earth) is very pronounced over the first 60-70 cm. A portion of the fine particles has, nevertheless, "plugged" the very rubefied B horizon. Due to the interacting effects of hydromorphism and acidity, the B horizon has acquired a characteristic variegated aspect (white spots resulting from "derubefaction" and secondary mobilization of reduced iron). These acid soils, having undergone prolonged development, nearly always contain kaolinite. However, in the present profile, 2/1 clays (illite) predominate.

REFERENCE: *Soil Map of France (1:100,000),* Montpellier sheet, Compagnie nationale du bas Rhône-Languedoc, Nîmes, 1974.

XVI₅: ELUVIATED FERSIALITIC RED SOIL (with ca horizon)

(Sol rouge fersiallitique lessivé)

F.A.O.: *Calcic Luvisol;* U.S.: *Petrocalcic Rhodoxeralf*

Location: Along the Ciudad Real to Valdepenas road, Spain.
Topography: Level site at the bottom of a depression surrounded by basalt hills. Elevation approx. 600 m.
Parent material: Fine colluvium of weathered basaltic material (plus some quartzous elements originating from quartzite).
Climate: Semi-arid Mediterranean. P. 500 mm; M.T. 14°C.
Vegetation: Olive tree plantation in the *Quercus ilex* and *Quercus lusitanica* forest climax zone.

Profile Description

A_p (0-30 cm):	Yellowish-red (5 YR 4/6) sandy clay loam; weak crumb structure, well aerated; abrupt boundary.
B_t (30-65 cm):	Dusky red (10 R 3/4) clay; strong prismatic structure; red, shiny clay skins, vertically oriented; abrupt boundary.
C_ca (65-90 cm):	Hard calcareous crust (the top of which can be seen in the photograph).
C:	Colluvium of weathered basaltic material with some quartzous elements.

Geochemical and Biochemical Properties

Particle-size distribution: The texture of this soil results from the twofold origin of the parent material: clay-size particles from basalt, sand from quartzite. Clay content is 27% in A_p and 53% in B_t with a translocation index of approximately 1/2. In addition to 2/1 clay minerals (illite and montmorillonite), a large proportion of kaolinite is present and indicates that the weatherable material (basalt) is low in silica.
Exchange complex: Due to the presence of kaolinite, the exchange capacity is relatively low for a Fersialitic soil (11 meq/100 g in A_p and 18 meq/100 g in B_t). The exchange complex is practically *base-saturated* with calcium and magnesium [S/T is 96% in B_t; pH (water) is 7.7].
Biochemistry: The humus has been greatly modified by cultivation. The A_p horizon contains 1.2% organic matter with a C/N ratio of 8. Under the forest climax, both the organic matter content and the C/N ratio would have been much higher.
Iron oxides: High free iron content (about 3% in B_t representing 68% of total iron). The translocation index for iron is comparable to that for clay.

Genesis. This soil displays most characteristics of an Eluviated Fersialitic Red soil developed from noncalcareous parent material: in particular, intense rubefaction by dehydration of iron oxides, which are mainly in the amorphous state and form films around the clay platelets, and strong eluviation of clay particles (together with bound iron) which occurs despite saturation of the exchange complex with calcium and magnesium. The calcareous crust is not a characteristic feature of Fersialitic soils (contrary to *Reddish Chestnut soils,* Chapter IV). When it does form, it is usually relatively deep and requires special conditions (of parent material or topography) to be present. Here, it has most probably formed from a lateral supply of calcium deriving from upslope ground water. The clay minerals of this soil contain, in addition to micaceous minerals, a large proportion of kaolinite due to the deficiency of the basaltic parent material in silica.

REFERENCE: Albareda, J.M., *Conference on Mediterranean soils,* Center of Edaphology, Madrid, 1966.

XVI₆: TRUNCATED, ELUVIATED FERSIALITIC RED SOIL (with ca horizon)

(Sol rouge fersiallitique lessivé, tronqué)

F.A.O.: *Calcic Luvisol;* U.S.: *Typic Rhodoxeralf*

Location: Algarafe, along the Sevilla to Huelva road, Spain.
Topography: Plateau with gentle slope. Elevation about 150 m.
Parent material: Mio-Pliocene calcareous sandstone.
Climate: Subhumid Mediterranean with pronounced dry season. P. 650 mm; M.T. 19°C (at Sevilla).
Vegetation: *Quercus ilex* and *Pinus pinea* forest with underbrush of *Quercus coccifera.*

Profile Description

A_1A_2 (0-10 cm):   Yellowish-red (5 YR 4/8) sandy loam, eroded humic horizon; single-grained structure, loose, aerated; few roots; gradual boundary.

A/B (10-25 cm):   Red (2.5 YR 4/6) sandy clay loam; blocky to subangular blocky structure; friable; gradual boundary.

B_t (25-70 cm):   Red (2.5 YR 5/6) sandy clay; coarse blocky to prismatic structure, less well expressed with depth; typical red clay skins; very hard when dry; abrupt boundary.

C_{ca}:   White tufaceous calcareous crust; nodular at places; grading into calcareous sandstone.

Geochemical and Biochemical Properties

Particle-size distribution: The solum consists of slightly clayey fine sand (70% sand) and is decarbonated in the upper 40 cm. Clay is eluviated and shows a very clear maximum (36%) in the upper part of B_t, with a translocation index of about 1/2. A comparison of the thickness of the A_1A_2 horizons versus the B_t horizon immediately suggests that the eluvial horizons have been truncated. Clay minerals consist of approximately 75% 2/1 clays and 25% kaolinite. As the former have migrated preferentially, there is a relative accumulation of kaolinite in the upper horizons.

Exchange complex: Exchange capacity is as expected considering the amount and nature of clay minerals (50 meq/100 g of *clay*). S/T is close to 100% at the surface and 85% in B_t (pH 6.6). Traces of non-active calcium carbonate at the base of B_t; 58% in C_{ca}.

Biochemistry: High organic matter content in A_1A_2 despite erosion (5.8% at the soil surface). C/N ratio of 15 in this horizon, but less than 10 in B_t.

Free iron: Fairly low free iron content (0.25% in A_2 and 0.70% in B_t). The free iron/ total iron ratio is abnormally low for a Fersialitic soil (about 0.30 in B_t). It increases towards the surface to 0.58 in A_1A_2. This provides an indication that this profile is *still young and incompletely mature with respect to weathering.*

Genesis. The fersialitic development of this soil has been accelerated by the calcareous sandstone nature of the parent material. Decarbonation of such a pervious material proceeds rapidly under the influence of the forest humus. The sandy loam residue is then subjected to marked contrasts in internal soil climate which accelerate rubefaction. Rubefaction takes place in a material that has not yet undergone complete geochemical weathering. At depth, a "crust" forms by precipitation of calcium bicarbonate leached from the surface. Because of the slope, a *global erosion* of the upper horizons replaces the normally observed "depletion," which, in fact, is really selective erosion.

REFERENCE: Albareda, J.M., *Conference on Mediterranean soils,* Center of Edaphology, Madrid, 1966.

PLATE XVI

FERSIALITIC SOILS

CHAPTER IX

FERRUGINOUS AND FERRALITIC SOILS

The soils of warm tropical and equatorial climates are separated into two distinct soil classes in the French system of soil classification; in reality, however, they possess common features linked to their deep and intense weathering, which occurs in neutral media low in organic matter.

In sufficiently drained environments, silica and bases are released through weathering and are eliminated to a large extent from the profile, whereas iron and aluminum oxides are retained. The degree of weathering of primary minerals is much greater than under temperate climate. Clay minerals result almost entirely from "neoformation" and are of the kaolinite type.

To distinguish between "Tropical Ferruginous soils" and Ferralitic soils is often a delicate exercise, as there exists a continuous gradation of soils from one group to the other. Some well-developed Tropical Ferruginous soils come very close to Ferralitic soils due to very strong weathering of their primary minerals; conversely, some Ferralitic soils are devoid of gibbsite as are Tropical Ferruginous soils.

DEFINITIONS AND NOMENCLATURE

It is difficult to summarize the different systems of soil classification of the world, because each is based on different principles. Nevertheless, I will attempt to do so using the French, U.S., F.A.O., and Belgian soil taxonomies.

Most of these use as a fundamental criterion the *extent of the loss of combined silica, which increases from the driest to the most humid climates*. The weathering complex becomes more and more depleted of silica, and clay minerals are found in the following sequence: (i) illite (vermiculite) + kaolinite; (ii) kaolinite; (iii) kaolinite + gibbsite; and (iv) gibbsite.

Ferralitic soils

Their main characteristics are: a complete weathering of primary minerals (except quartz); clay minerals of the kaolinite type only; low exchange capacity (less than 16 meq/100 g of clay in the U.S. soil taxonomy); and no eluviation of clay which resists to dispersion. According to the extent of loss of silica, the following groups may be distinguished:

137

Ferralites: Gibbsite predominates, and there is little kaolinite (NOTE: I will use the term *alite* in hydromorphic environments where iron is mobilized through reduction).

Ferralitic soils with gibbsite: Kaolinite predominates over gibbsite.

Ferralitic soils with kaolinite: There is very little or no gibbsite.

Ferrisols

These are transitional soils which are given a special grouping in the Belgian soil taxonomy and by the F.A.O. (*Nitosols*). Weathering of primary minerals is incomplete and some water-dispersible 2/1 clay minerals may be moderately eluviated. A barely distinguishable B_t horizon forms with weakly expressed cutans and a gradual boundary with either the A or C horizon (Jamagne, 1963).

Tropical Ferruginous soils

Although they are highly weathered, these soils still contain the most resistant primary minerals. Small amounts of 2/1 clay minerals remain and are likely to migrate. Such soils are usually eluviated and are characterized by a well-expressed argillic B_t horizon.

The French system of soil classification, which has used African soils as models (thus under *dry* tropical climate), describes these soils as being slightly desaturated, whereas the U.S. system, based on American models under more humid climate, classifies them with the strongly desaturated order of *Ultisols*. In fact, both types occur depending on climate.

Actually, base saturation is an inadequate criterion to define the subclass level. I will adopt the view of Sys (1967) that the exchange capacity of clays (ranging between 16 and 25 meq/100 g) is a better yardstick and allows us to distinguish between the two groups. These can then be separated according to the base status of clays as follows: (i) slightly desaturated Tropical Ferruginous soils (S/T above 50%)—Alfisols (U.S.) and Luvisols (F.A.O.) under dry climate; and (ii) desaturated Tropical Ferruginous soils (S/T less than 50%)—Ultisols (U.S.) and Acrisols (F.A.O.) under humid climate.

GENESIS

As the fundamental criterion of classification is the loss of combined silica during the weathering process, it is important to specify the relationships that exist between this criterion and ecological factors. The extent to which silica is lost depends primarily on the time factor (length of development), then on the general climatic factors, and, finally, on site factors (parent material and topography). In order to study the effects of the first two factors, it will be necessary to consider sufficiently *drained* (leaching) environments.

Time factor (Icole, 1973)

The time factor can be emphasized by considering the evolution of alluvial deposits of differing ages, in a region with a sufficiently humid climate that soil formation may proceed until ferralitization takes place, but not so humid that development progresses too rapidly. It is then possible to distinguish the main steps of the evolution sequence, which, under humid climate with contrasting seasons, derive from the following four processes:

Process 1. Weathering of primary minerals.

Process 2. Loss of combined silica and decrease in the exchange capacity.

Process 3. Increasing desaturation of the exchange complex.

Process 4. Reduction of the "activity" of clay minerals and of leaching.

Eutrophic Brown soil (initial stage)	→Slightly desaturated Tropical Ferruginous soil	→Desaturated Tropical Ferruginous soil	→Ferrisol	→Ferralitic soil

Increasing degree of development: Processes 1-4

General climate and climatic vegetation (climatic climax)

In plains, the degree of weathering of, and the loss of silica from a material of given age are linked to the mean annual precipitation of the region. "Zonation" from more to less ferralitized soils occurs from south to north in Central Africa (although it is incomplete as the "humid tropical" zone is only partially represented). Such a soil zonation also exists on the American Continent, in Colombia for instance, where it is more complete, but takes the form of "climatic compartments," rather than being delineated by latitudes. The compartments are tied to relief barriers which block the moist winds.

At high elevations, *vegetation zones*, as well defined as under temperate climate, occur along a gradient of decreasing *mean temperature*. Table 11 presents an outline of soil zonation with latitude (plains) and with altitude (elevation zones).

Soil zonation in plains

Very humid equatorial climate. Example: Pacific Coast of Colombia; P. 8-10 m; no dry season; tropical rainforest. Intense ferralitization is observed and iron is mobilized through reduction and leached from the drained soil zones.

Humid equatorial climate. Example: Amazonia, Colombia; P. 3-4 m; tropical rainforest. Ferralitic soils are colored with iron (in "ocher" or "red" depending on the intensity of the short periods of desiccation of the profile).

Humid tropical climate. Example: Llanos Plain, Colombia; P. 2-3 m; short dry season. All transitions exist with the equatorial climate depending upon the length of the dry season. Vegetation consists of a "mesophytic" forest which, in Africa, has often been cleared and replaced by a secondary savanna with grasses. As soil formation is slower, the "age of the material" plays an essential role. Accordingly, desaturated Tropical Ferruginous soils, Ferrisols, or even Ferralitic soils (Ivory Coast) may be found.

Dry tropical climate. "Sudan" type in Central Africa; P. 1-1.5 m; marked dry season. The primary savanna consists of more or less xerophytic shrub formations (thorn scrub or Combretaceae). Plant formations in which grasses predominate are often secondary. This is the region of slightly desaturated Tropical Ferruginous soils.

Elevation zones (Central Cordillera, Colombia). As the mean temperature decreases with altitude, there is a reduction in potential evapotranspiration accompanied by a correlative increase in soil water content. These two factors have a twofold effect on soil formation: (i) weathering decreases with altitude and easily dispersible 2/1 clay minerals remain, clay "eluviation" increases; (ii) the organic matter is decomposed more slowly and the profiles become more humic. The latter characteristic is frequently accompanied by hydromorphism and cryptopodzolization at the surface with the appearance of *secondary gibbsite* (resulting from the weathering of primary minerals, and even of kaolinite, at the base of the humic horizons).

Site factors

The effects of the two site factors, *parent material* and *topography*, are intimately bound to each other. They are the cause of relatively localized soil-vegetation equilibria, which have been designated as "site climaxes." *In certain poorly drained areas, enriched with laterally supplied silica and divalent cations, ferralitization is locally replaced by a vertic soil formation process with "neoformation" of montmorillonite. This constitutes a typical site climax* (Perraud, 1971).

The role played by local "drainage conditions" (i.e., the more or less ready removal of the soil solution) is of paramount importance. Three situations will be considered (Paquet, 1969): well-drained (or leaching) environments, lowest in silica, on summits or at the top of slopes; semiconfined environments with medium silica contents, half-way down slopes; and confined environments at the foot of slopes and often characterized by the predominance of 2/1 clay minerals of neoformation (montmorillonite).

However, topographic soil sequences or "catenas" differ greatly depending on the composition of the parent material. In this respect, *basic* crystalline rock (low in quartz and combined silica) will be compared with *acid* crystalline rock (with quartz).

On basic rock, there is a strong opposition between the soil forming processes that take place on well-drained summits and those occurring in confined environments at the foot of slopes. From top to bottom, *Ferralites* are followed in succession by *Ferralitic soils*, then by *Vertic Eutrophic Brown soils*. (These are different from "incipient" Eutrophic Brown soils which are young and incompletely weathered.) Under drier climate, even true black Vertisols form in these confined media (upward movement of montmorillonite; Paquet, 1969; Bocquier, 1973).

On acid rock and because of large reserves of silica (even quartz has an effect, despite its very slow weathering), contrasts between top and bottom of slopes are attenuated. The soil environment is too acid for the neoformation of montmorillonite; only kaolinite can form. Thus, weathering is of the "ferruginous" or "ferralitic" type. Loss of silica is, nevertheless, more important in well-drained areas (ferralitization with presence of "gibbsite") than in poorly drained areas (formation of Ferruginous soils or Ferralitic soils without gibbsite; Segalen, 1973).

In addition, acidity appears rapidly in the upper horizons of soils developed on these parent materials and promotes the mobilization of iron, which is then transported downslope in the ground water. Special horizons form, such as *plinthite*—a hydromorphic horizon locally enriched with iron (Profile XVIII$_1$)—or even ironstone hardpans, formed by lowering of the water table (Profile XVIII$_4$).

APPENDIX: ECOLOGY OF IRONSTONE HARDPAN FORMATION

Under particular ecological conditions (often combined with human intervention), induration may take place and lead to the formation of *ironstone hardpans*. These result from the *incrustation of one or more horizons (ferruginous or ferralitic) with sesquioxides (especially iron) which crystallize and harden at high ambient temperatures.*

Three examples of hardpan formation will be discussed, based on the results of D'Hoore (1955), Maignien (1958), and Van Wambeke (1971). Formation processes will be distinguished according to the *mode of accumulation* of sesquioxides ("absolute" accumulation when cementation agents are supplied directly by ground water or "relative" accumulation when other constituents are eliminated selectively) and according to the rate of formation.

1. Rapidly forming hardpans resulting from absolute accumulation. Iron, supplied by the lateral flow of upslope ground water, accumulates in low points, often in the form of plinthite (Profile XVIII$_1$). Desiccation following an overall decline of the water table induces hardening of the plinthite into petroplinthite (Profile XVIII$_4$).

2. Slowly forming hardpans resulting from mixed accumulation. During the "climatic" forested phase, a relative accumulation of iron occurs first by removal of silica and bases. This is followed by a concentration of iron oxides by absolute accumulation in low points or at breaks of slope (Maignien, 1958). Clearing of the forest or its destruction accentuates the process and induces the lateral and vertical extension of the indurated "nuclei." Such hardpans, so-called *plateau* or *erosion* hardpans (the loose upper soil horizons are progressively eliminated), are very slow to form and are characteristic of the *secondary savanna* under contrasting climate in Africa (Guinean region). This process is often accompanied by an *inversion of relief* and the formation of extensive "platforms" of erosion.

3. Hardpans resulting from relative accumulation. The extreme evolution of Ferralites on ultrabasic rock, thus very low in silica, leads to the total elimination of all constituents other than sesquioxides. Hardpans can then form, but they are relatively rare and are found only in areas where the degradation of the protective forest has enabled hardening to occur by local crystallization of oxides.

NOTE RELATING TO THE PRESENTATION OF SOIL ENTRIES: Due to the considerable thickness of the profiles, horizon designations must be adapted to tropical soils. Only A$_1$, A$_2$, and B horizons are recognizable in the upper portion of the profiles. Below, very thick zones of alteration can be distinguished. These are often partially hydromorphic and variegated (*saprolite* or *mottled zone*, designated M.Z. in the profile descriptions).

Furthermore, because the weathering process and the type of clay minerals are so important, a special paragraph entitled "weathering products" has been inserted in the Geochemical and Biochemical Properties section. Iron and aluminum contents are expressed as percent *oxides* (Fe_2O_3 for goethite and hematite; Al_2O_3 for gibbsite).

Table 11.   Ecology of Equatorial and Tropical Soils[a]

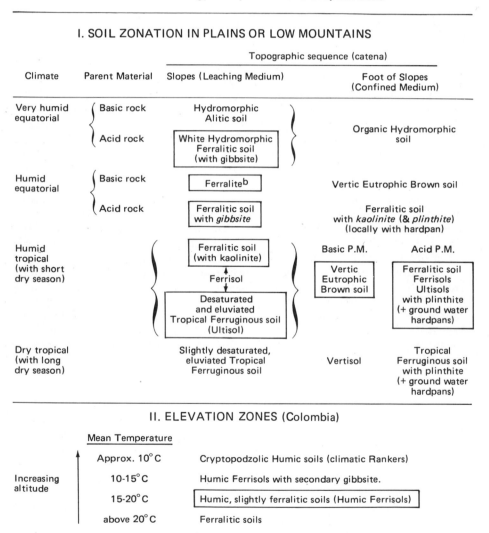

I. SOIL ZONATION IN PLAINS OR LOW MOUNTAINS

Climate	Parent Material	Topographic sequence (catena)	
		Slopes (Leaching Medium)	Foot of Slopes (Confined Medium)
Very humid equatorial	Basic rock	Hydromorphic Alitic soil	Organic Hydromorphic soil
	Acid rock	White Hydromorphic Ferralitic soil (with gibbsite)	
Humid equatorial	Basic rock	Ferralite[b]	Vertic Eutrophic Brown soil
	Acid rock	Ferralitic soil with *gibbsite*	Ferralitic soil with *kaolinite* (& *plinthite*) (locally with hardpan)
Humid tropical (with short dry season)		Ferralitic soil (with kaolinite) ↑ Ferrisol ↓ Desaturated and eluviated Tropical Ferruginous soil (Ultisol)	Basic P.M.: Vertic Eutrophic Brown soil Acid P.M.: Ferralitic soil Ferrisols Ultisols with plinthite (+ ground water hardpans)
Dry tropical (with long dry season)		Slightly desaturated, eluviated Tropical Ferruginous soil	Vertisol Tropical Ferruginous soil with plinthite (+ ground water hardpans)

II. ELEVATION ZONES (Colombia)

Increasing altitude	Mean Temperature	
↑	Approx. 10°C	Cryptopodzolic Humic soils (climatic Rankers)
	10-15°C	Humic Ferrisols with secondary gibbsite.
	15-20°C	Humic, slightly ferralitic soils (Humic Ferrisols)
	above 20°C	Ferralitic soils

[a]Adapted in part from Lamouroux (1972) and Bocquier (1973).
[b]The example studied (Profile XVII$_4$) displays some characteristics of a high-elevation Humic Ferrisol.

XVII₁: DESATURATED TROPICAL FERRUGINOUS SOIL
(Sol ferrugineux tropical désaturé)

F.A.O.: *Orthic Acrisol*; U.S.: *Typic Tropudult*

Location: Las Brisas, near Villavicensio (Llanos), Meta, Colombia.
Topography: Subhorizontal alluvial cone, piedmont of the Eastern Cordillera.
Elevation 400 m.
Parent material: Colluvium and alluvium, crystalline materials.
Climate: Humid tropical. P. 2,500-3,000 mm (3-month dry season); M.T.
25°C.
Vegetation: Mesophytic forest; cleared area, cultivated to rice and meadow.

Profile Description

A₁ (0-22 cm):   Very dark brown (10 YR 2/2) humic sandy loam, angular crumb struc-
                  ture; slightly plastic; many fine roots; gradual boundary.
A₂ (22-65 cm):   Yellowish-brown (10 YR 5/6 to 5/8) loam; weak to massive structure;
                  slightly plastic; few roots; gradual boundary.
B_{t1} (65-105 cm):   Strong brown (7.5 YR 5/6) clay loam, slightly spotted with red; moder-
                  ate blocky structure, with weakly expressed shiny clay skins; few roots;
                  diffuse boundary.
B_{t2} (105-150 cm): Strong brown (7.5 YR 5/8) clay loam; same structure as above; plastic.

Geochemical and Biochemical Properties

Particle-size distribution: Sandy loam at the surface to clay loam at depth with 12%
clay in A₁, 24% in A₂, 34% in B_{t1}, and 36% in B_{t2}. Thus, clay accumulation is very gradual
and dispersed. Presence of weakly expressed argillans, yet noticeable to the unaided eye.

Weathering products: The clay-size fraction contains no gibbsite but is composed of
kaolinite and illite in equal amounts. Exchange capacity is 20 meq/100 g of *clay*, characteristic of Tropical Ferruginous soils.

Exchange complex: The profile is highly desaturated and very strongly acid. The
value of S ranges from 0.5 to 0.6 meq/100 g in all horizons. Under these conditions, S/T
amounts to 7.6% in B_t, but only to 4.2% in A₁, where the exchange capacity exceeds
10 meq/100 g because of the presence of organic matter. pH (water) is 4.3 in A₁ and 4.8 in B_t.

Biochemical properties: Organic matter content is low throughout, except in A₁
(2.5%). The humus is dark-colored due to a certain degree of climatic maturation linked to
the dry season.

Genesis. The presence of an argillic horizon indicates that this is a typical
desaturated and eluviated Tropical Ferruginous soil (Ultisol). However, the dispersed accumulation of clay brings this soil close to a Ferrisol (Nitosol). An
exchange capacity of 20 meq/100 g of clay in B_t is characteristic of this category of
soils (Sys, 1967). The relative abundance of inherited and still nonweathered illite
is a sign of a fairly young profile. related to the age of the alluvial parent
material. It is interesting to compare this profile to Profile XVII₃, which is also
located in the Llanos Plain but much farther from the mountains, thus on older
parent material. The latter is a Ferralitic soil from which illite has disappeared.
This fact can be attributed neither to bioclimatic factors nor to the nature of the
parent material—they are similar—but to a longer evolution, thus a longer period
of weathering. Although it is less weathered, the present profile is, however,
more eluviated than Profile XVII₃, because illite is more "mobile" than kaolinite.

An example of a *slightly desaturated Tropical Ferruginous soil* would be
Profile XV₅, except for the fact that, as I have mentioned, it has undergone
planosolization at the surface.

REFERENCE: Faivre, P., and J. Rivillas, Codazzi Institute, Bogota, Colombia,
1971.

XVII$_2$: VERTIC EUTROPHIC BROWN SOIL

(Sol brun eutrophe vertique)

F.A.O.: *Chromic Vertisol*; U.S.: *Typic Chromoxerert*

Location: Along the Seguela to Sifié road, 10 km from Seguela, Ivory Coast.
Topography: Foot of slope, with poor drainage. Elevation 300 m.
Parent material: Schist with amphibole.
Climate: Humid tropical with a 6-month dry season. P. 1,400 mm; M.T. 23°C.
Vegetation: Savanna with sparse trees (anthropic secondary savanna).

Profile Description

A$_1$ (0-15 cm):   Brown (7.5 YR 5/4) clay loam to clay; medium crumb structure; firm; common fine and medium roots; gradual boundary.

(B) (15-50 cm):   Pinkish-gray (7.5 YR 6/2) clay with reddish-yellow (7.5 YR 6/6) diffuse spots; coarse prismatic structure, with diagonal slickensides; few roots; gradual boundary.

(B)$_{ca}$ (50-110 cm):   Pinkish-gray clay; same structure as above with slickensides; reddish-yellow spots increase in number with depth; presence of white calcareous nodules; gradual boundary with the C horizon.

Geochemical and Biochemical Properties

Particle-size distribution: Texture is characteristic of Vertic soils. High clay content, especially of swelling clays (40% clay in A$_1$ and 58% at 100 cm). The decrease in clay content in the surface horizon appears to be due to depletion, as shown by the increase towards the soil surface of quartz sand (close to 30% in A$_1$ versus 12% at 100 cm).

Weathering products: In a confined medium high in bases, neoformation of clay minerals produces a large proportion of swelling clays: montmorillonite exceeds kaolinite. This is reflected in a high exchange capacity of 25 meq/100 g in A and (B), or 60 meq/100 g of *clay*, which is an intermediate value between the exchange capacity of Tropical Ferruginous soils and that of typical Vertisols. Free iron content is fairly high [4.7% in (B)], but the free iron/total iron ratio barely exceeds 0.5, which is evidence of a lower degree of weathering than in Ferruginous soils (presence of ferriferous montmorillonite; Paquet, 1969).

Exchange complex: The exchange complex is obviously base-saturated with Ca^{2+} and Mg^{2+} in approximately equal amounts. A slight "supersaturation" in calcium, resulting from lateral input, leads to the precipitation of excess calcium as nodules in (B)$_{ca}$.

Organic matter: O.M. content is 2.3% in A$_1$ with a C/N ratio of 10. There is very little organic matter in (B); it does not darken the horizon, in opposition to typical Vertisols.

Genesis. The well-characterized vertic evolution of this soil is due in part to a general climatic factor—a dry season—and also to "site factors"—topographic position and parent material rich in weatherable minerals, releasing large amounts of Ca^{2+} and Mg^{2+}. Bases and silica are released uphill, then migrate laterally and accumulate at the foot of the slope. Such ecological conditions promote the neoformation of montmorillonite. However, this soil differs from the typical Vertisol; its vertic features are less pronounced: clay minerals contain a considerable proportion of kaolinite, the exchange capacity is lower than in Vertisols, and "stabilized" organic matter penetrates less deeply. Moreover, this soil should not be confused with two other kinds of Eutrophic Brown soils, i.e., Temperate Eutrophic Brown soils whose weathering products are entirely different ("transformed" clay minerals predominate) and Tropical Eutrophic Brown soils, which are young soils formed, contrary to this soil, on well-drained, often eroded slopes and constitute the incipient phase in the evolution of Ferruginous soils.

REFERENCE: Latham, 1970.

XVII₃: FERRALITIC SOIL (with kaolinite)

(Sol ferrallitique à kaolinite)

F.A.O.: *Orthic Ferralsol*; U.S.: *Typic Haplorthox*

Location: 1,500 m northwest of Las Gaviotas, Llanos, Vichada, Colombia.
Topography: Subhorizontal (1-3% slope). Elevation 200 m.
Parent material: Alluvium and colluvium of old crystalline materials.
Climate: Humid tropical, 3-month dry season. P. 2,500 mm; M.T. 27°C.
Vegetation: Savanna with grasses.

Profile Description

A₁ (0-20 cm): Dark brown (10 YR 3/3) humic clay loam, with scattered gray or red spots; angular crumb structure; common grass roots; intense biological activity; gradual boundary.

(B)₁ (20-60 cm): Reddish-yellow (7.5 YR 6/8) clay; weak blocky structure; porous, friable, slightly plastic; few fine roots; gradual boundary.

(B)₂ (60-100 cm): Yellowish-red (5 YR 5/8) clay; weak blocky structure, with localized shiny "stress cutans" (hardened areas alternating with softer ones); small red concretions; diffuse boundary.

(B)₃ (100-160 cm): Red (2.5 YR 4/8) clay; massive structure; friable, slightly plastic; some ocherous spots.

Geochemical and Biochemical Properties

Particle-size distribution: Except for a moderate "depletion" of clay in the humic horizon (38% clay), texture is homogeneous with 56% clay, 30% silt, and 14% sand. Clay illuviation is neither morphologically nor analytically discernible in the (B) horizon.

Weathering products: The weathering products consist of kaolinite and crystallized iron oxides (6.5% Fe_2O_3). *There is no gibbsite.* Iron content is constant throughout the profile, except in the depleted A₁ horizon.

Exchange complex: Exchange capacity is low (6.2-6.4 meq/100 g or 12 meq/100 g of *clay*), which places this soil in the "Oxisol" order. The value of S is extremely low (0.6-0.5 meq/100 g), S/T is about 7%, and pH (water) varies from 5 at the surface to 6 at depth.

Biochemical properties: Very low organic matter content (4% in A₁) with a C/N ratio of 15, indicating a weak humification and a rapid biodegradation of humus substances.

Genesis. This soil displays all the essential features of Ferralitic soils. Weathering of primary minerals excepting quartz is practically complete, clay minerals consist exclusively of kaolinite of neoformation, and there is no clay illuviation. The mechanical depletion at the surface is a constant of warm climates and is not accompanied by illuviation in the (B) horizon. Exchange capacity is lower than 16 meq/100 g of clay and is characteristic of "Oxisols" in the U.S. soil taxonomy. Thus, this soil must be considered "ferralitic." However, the almost total lack of free Al_2O_3 (gibbsite) differentiates it from the more mature Ferralitic soils with gibbsite. Here, ferralitization is moderate and proceeds slowly under a moderately humid climate, on old alluvial deposits. Desaturated Tropical Ferruginous soils (Profile XVII₁) are found on more recent parent materials.

NOTE. In wet environments, weathering at the surface is influenced by organic matter and produces brownish-yellow goethite. Conversely, at depth and in drier environments, weathering produces red hematite as crystallization is more rapid in the absence of organic matter (Schwertmann et al., 1974).

REFERENCE: Faivre, P., Codazzi Institute, Bogota, Colombia, 1972. (Photo by P. Faivre.)

XVII₄: FERRALITIC SOIL (with gibbsite)
(Sol ferrallitique à gibbsite)

F.A.O.: *Xanthic Ferralsol*; **U.S.:** *Typic Haplorthox*

Location: Along the Pasto to Mocoa road, Amazonia, Putumayo, Colombia.
Topography: Very gentle slope (2-3%). Elevation 800 m.
Parent material: Arenaceous weathering products of reworked crystalline rock (dejection cone).
Climate: Humid equatorial. P. 4,000 mm; M.T. 26°C.
Vegetation: Cleared rainforest cropped to sugar cane.

Profile Description

A_p (0-18 cm):   Brownish-yellow (10 YR 6/6) slightly humic clay; weak blocky structure, with shiny ped surfaces; porous, slightly plastic.

(B) (18-140 cm):   Reddish-yellow (7.5 YR 6/8) clay, grading to light red (2.5 YR 6/8) with reddish-yellow (5 YR 6/8) spots; weak coarse blocky structure, with shiny stress-cutans at places; slightly sticky, plastic, porous; gradual boundary.

C (140-230 cm):   Red (2.5 YR 5/8) clay with yellow (10 YR 7/6) spots; massive structure; slightly sticky, plastic; strongly weathered rock fragments at depth (muscovite and quartz are still visible).

Geochemical and Biochemical Properties

Particle-size distribution: Except for a typical "depletion" in A_p, clay content is very high [62% in (B)]. It decreases slightly in C (50%) where weathering is not entirely over.

Weathering products: The clay-size fraction is composed of about two-thirds kaolinite and one-third gibbsite and goethite. There are no micaceous clay minerals.

Exchange complex: The exchange capacity is somewhat in excess of the accepted norm for Oxisols (18 meq/100 g of *clay*). This may be explained by the presence of negative charges on the silt-size particles (kaolinite in large crystals). The value of S is very low (0.5 meq/100 g); S/T does not exceed 3.8% in (B) (pH 5.3).

Organic matter: Low organic matter content in A_p (less than 5%). Biodegradation is very rapid as is usually the case in Ferralitic soils formed in plains under very humid climate.

Genesis. Despite having an exchange capacity slightly above the norm (explained by the presence of kaolinite in the silt fraction), this soil exhibits all the usual features of Ferralitic soils. Weathering of the primary minerals is complete (excepting quartz), kaolinite clearly predominates, but sesquioxides (especially gibbsite) represent a nonnegligible fraction of the weathering products. Thus, its place in the evolution sequence is intermediate between typical "Ferralites" (Profile XVII₆) and Ferralitic soils with kaolinite but devoid of gibbsite (Profile XVII₃). The overall color of the profile is reddish-yellow, especially in the upper portion, and changes to red at depth. This indicates the predominance of hydrated goethite over hematite and could be due to a wetter soil climate than that of strong red Ferralitic soils with hematite, but much less humid than that of white Ferralitic soils (Profile XVII₅). Ferralitic soils formed in plains contrast with those formed at high elevation in that, in the former, mineralization of the organic matter is very rapid.

REFERENCE: Faivre, P., Codazzi Institute, Bogota, Colombia, 1972. (Photo by P. Faivre.)

XVII₅: HYDROMORPHIC FERRALITIC SOIL

(Sol ferrallitique hydromorphe)

F.A.O.: *Humic Ferralsol*; U.S.: *Umbraquox*

Location: Pacific Coast, Cauca, Colombia.
Topography: Coastal plain at the top of a hill. Elevation 30 m.
Parent material: Weathered Tertiary lutite.
Climate: Very humid equatorial. P. 8,000 mm; M.T. 26°C.
Vegetation: Tropical rainforest (very hydrophilous vegetation).

Profile Description

A_0 (18-0 cm):   Dark brown (7.5 YR 3/2) raw humus and litter; slow and incomplete decomposition; platy structure, white mycelium filaments; abrupt boundary.

$(B)_{1g}$ (0-20 cm):   Gray to light gray (7.5 YR 6/0) clay, with very small reddish-yellow (7.5 YR 7/6) mottles; massive structure, sticky; permanently water-saturated; common roots; diffuse boundary.

$(B)_{2g}$ (20-130 cm):   White (7.5 YR 8/0) clay; massive structure, sticky; permanently water-saturated; *friable*, weathered quartz pebbles.

Geochemical and Biochemical Properties

Particle-size distribution: Homogeneous texture throughout the profile; very high in clay (64-66%).

Weathering products: The clay-size fraction consists almost exclusively of kaolinite; some gibbsite is present, however. Iron oxide content is very low (less than 1%). Quartz pebbles, even large ones, are weathered and friable.

Exchange complex: The exchange capacity of the mineral horizons is 12 meq/100 g or 18 meq/100 g of *clay*, which is slightly above the Oxisol level, but can be explained by the abundance of silt-size kaolinite. Therefore, this is indeed an Oxisol. The exchange complex is strongly desaturated. The value of S is very low in the mineral horizons (0.5 meq/100 g), but exceeds 20 meq/100 g in A_0 because of the considerable effect of the biogeochemical cycle. S/T is 4 to 4.5% in $(B)_g$, but 24% in A_0 [pH 4.8 in A_0, 4.5 in $(B)_g$].

Biochemical properties: A_0 contains 36% organic matter with a C/N ratio of 18. This horizon constitutes a true tropical *mor* with a slow decomposition rate. There is still 1.3% organic matter in $(B)_{1g}$, mainly as fulvic acids.

Genesis. The genesis of this profile is akin to the genesis of Profile XVII₄ (Ferralitic soil with gibbsite), but with two important distinctions: (i) iron is almost totally eliminated from the profile; (ii) the weathering products contain very little gibbsite with respect to kaolinite. The first characteristic must be attributed to the extremely humid climate together with the influence of the soluble organic compounds that form in the A_0 horizon. Practically all the iron is reduced and solubilized in this very strongly acid medium. It migrates down the slope and accumulates at its bottom (presence of plinthite at the foot of the slope). The second characteristic can probably be explained by the nature of the sedimentary clay parent material. Although it may be intense, ferralitic weathering of argillaceous materials releases less free alumina than that of crystalline rocks. The 2/1 lattice clays lose part of their silica and are transformed into kaolinite, which is stable in a semiconfined medium. Furthermore, a large proportion of fine quartz grains is dissolved and eliminated; even large quartz pebbles are "comminuted" by weathering.

REFERENCE: Faivre, P., Codazzi Institute, Bogota, Colombia, 1972. (Photo by P. Faivre.)

XVII6: HUMIC FERRALITE
(Ferrallite humifère)
F.A.O.: *Humic Ferralsol*; U.S.: *Typic Gibbsihumox*

Location: Palni Hills, overlooking the Vattavadai Plain, India.

Topography: Summit at top of 30% slope, west exposure. Elevation 2,490 m.

Parent material: Enderlite with hypersthene and hornblende (crossed by quartz veins).

Climate: High elevation tropical with a 2-month dry season. P. 1,450 mm; M.T. approx. 14.5°C.

Vegetation: High elevation meadow with *Pollinia phaeotrix* and *Arundinella* sp. (replacing the burnt montane forest).

Profile Description

A_0A_1 (0-10 cm): Very dark brown (10 YR 2/2) very humic horizon; fine crumb structure, aerated; many roots.

A_1(B) (10-30 cm): Heterogeneous horizon; large zones of black humic crumbs forming an incipient "sombric" horizon, alternating with reddish-yellow (5 YR 6/8) mineral fragments undergoing weathering; clear boundary.

(B)$_1$ (30-60 cm): Reddish-yellow (7.5 YR 6/8) sandy clay loam, with many weathered rock fragments; fine crumb structure.

(B)$_2$ (60-100 cm): Yellowish-red (5 YR 5/8) sandy clay loam matrix around large exfoliating rock masses.

(B)C: Stratification of more or less weathered horizontal slabs and sandy loam layers.

Geochemical and Biochemical Properties

Particle-size distribution: It is not very significant in the case of this still "young" profile, because the parent material is so heterogeneous. The fine earth is a sandy clay loam with 50-60% sand and 28-32% clay. Clay content decreases slightly with depth.

Weathering products: Their composition was determined by the geochemical "balance" method (Souchier, 1971). In A_1, the crystalline fraction represents 15% of which two-thirds are gibbsite and one-third kaolinite with traces of quartz; the amorphous fraction represents 17% of which two-thirds are cryptocrystalline iron oxides. In (B), the crystalline fraction represents 11% of which three-quarters are gibbsite and one-quarter kaolinite, whereas amorphous materials amount to 17% with two-thirds iron. Note the presence of gibbsite in the silt and sand fractions.

Exchange complex: The exchange capacity is high at the surface because of the presence of organic matter [46 meq/100 g in A_0A_1 and 19 meq/100 g in A_1(B)]. In the mineral horizons, it does not exceed 2 to 3 meq/100 g and appears to be also essentially due to the small organic fraction. The exchange complex is strongly desaturated (S/T is 6% in the A and B horizons) and pH is about 5.3.

Biochemical properties: High organic matter content in the upper horizons [28% in A_0A_1 with a C/N ratio of 26 and 10% in A_1(B) with a C/N ratio of 16]. Small amounts of organic matter exist in (B)$_1$ and (B)$_2$.

Genesis. This soil has formed at an excessively "drained" site due to the proximity of the steep slope, on a basic parent material low in quartz. Because of the intense extraction of combined silica and bases, it evolves rapidly towards a *Ferralite* in which the weathering products are characterized by the predominance of sesquioxides (about 80%, mainly gibbsite) over kaolinite. This soil displays certain peculiarities related to its "young" age (a few percent of primary minerals remain) and to the heterogeneity of the parent material. The profile still contains incompletely weathered fragments and an abnormally abundant sand matrix consisting of residual quartz grains. Furthermore, the high elevation of the site, together with a decrease in mean annual temperature combine to slow the rate of decomposition of the organic matter and give the profile a very humic aspect not unlike that of high elevation Humic Ferrisols (Profile XVIII$_2$).

REFERENCE: Troy, J.P., Thesis (in preparation), University of Nancy I. (Photo by J.P. Troy.)

PLATE XVII
FERRUGINOUS AND FERRALITIC SOILS

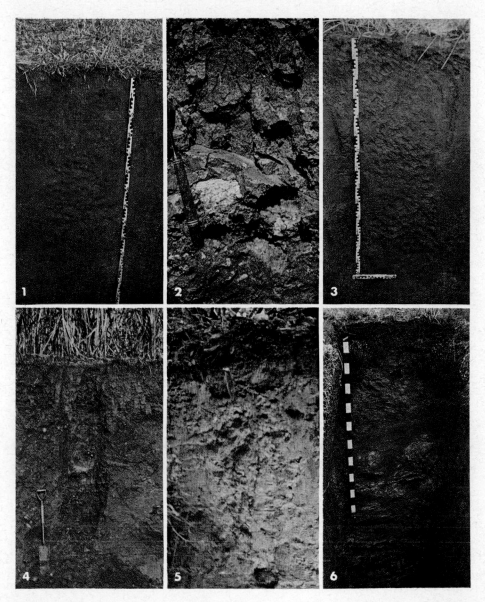

XVIII₁: FERRISOL (with plinthite)

(Ferrisol à plinthite)

F.A.O.: *Plinthic Nitosol*; U.S.: *Typic Plinthudult*

Location: 3.5 km from La Rochela, Santander, Colombia.
Topography: Undulating hills; moderate slope. Elevation 270 m.
Parent material: Tertiary argillaceous schist.
Climate: Humid tropical to equatorial, with a 3-month semidry season. P. 2,500 mm; M.T. 29°C.
Vegetation: Partially cleared mesophytic forest; pasture with silk-cotton trees.

Profile Description

A_1A_2 (-20 cm): Dark yellowish-brown (10 YR 4/4) loam; angular crumb structure; firm, slightly plastic; common roots; clear boundary.

A/B (20-50 cm): Red (2.5 YR 4/6) clay, with some paler diffuse spots; strong blocky structure; firm, plastic; common roots; gradual boundary.

B_1 (50-90 cm): Dark red (10 R 3/6) clay, with vertically oriented yellow and gray streaks; blocky structure; sticky, very plastic; few, weakly expressed dispersed coatings; few roots, gradual boundary.

B_2 (90-150 cm): Dark red (10 R 3/6) clay in vertical bands alternating with light gray (10 YR 7/1), sometimes yellowish-brown bands (*plinthite*); massive structure; very sticky, very plastic.

Geochemical and Biochemical Properties

Particle-size distribution: The A_1A_2 horizon is markedly depleted of clay (24% clay); clay content increases gradually from 52% in A/B to 70% in B_2. Illuviated clay is very dispersed and does not form well-expressed cutans in B.

Weathering products: Not studied in detail. However, the exchange capacity exceeds that of an Oxisol (20 meq/100 g of *clay* in A/B and B_1; 26 meq/100 g in B_2). This would suggest that clay minerals are not exclusively kaolinite, but include a certain proportion of micaceous minerals, which increases with depth.

Exchange complex: The value of S is very low, but increases with depth (0.5 meq/100 g in A_1A_2 to 2.9 meq/100 g in B_2). S/T increases from 5% in A/B to 16% in B_2. Soil reaction is very strongly acid (pH 4.5 in A/B, 4.9 in B_2).

Organic matter and soil climate: Organic matter content is very low (about 2%) in A_1A_2 and A/B. The deep horizon has only capillary pores and is almost water-saturated during most of the year (except during the "semidry" season).

Genesis. The weathering in this profile is less complete than in "Oxisols," as illustrated by an exchange capacity above 16 meq/100 g of clay. Thus, in addition to kaolinite, a small proportion of micaceous clay minerals is present. These minerals have accumulated gradually and in a dispersed fashion, as indicated by an increase in the exchange capacity with depth. Although there is an argillic horizon, it is not clearly expressed. These are features of "Ferrisols" ("Nitosols" in the F.A.O. classification). Moreover, the B_2 horizon constitutes a "plinthite," very rich in free iron which is concentrated in large, diffuse streaks under the effect of prolonged water saturation. Contrary to "pseudogleys," the red color of iron indicates that seasonal re-oxidation occurs rapidly as it is not slowed by the presence of organic matter (Schwertmann et al., 1974). This horizon can harden following drainage and change into a "petroplinthite" (Profile XVIII₄).

REFERENCE: Faivre, P., Codazzi Institute, Bogota, Colombia. (Photo by P. Faivre.)

XVIII₂ & ₃: HIGH ELEVATION HUMIC FERRISOL
(Ferrisol humifère d'altitude)

F.A.O.: *Ferralic Cambisol*; U.S.: *Oxic Dystropept*

Location: Along the Gigante to Tres Esquinas road, Huila, Colombia.
Topography: Eastern Cordillera foothills, 7-12% slope. Elevation 1,300 m.
Parent material: Weathered granite and gneiss detrital material.
Climate: High elevation tropical with not very pronounced dry season. P.
1,300 mm; M.T. 19.3°C.
Vegetation: Partially cleared high elevation tropical forest; meadow.

Profile Description

A₁ (0-40 cm): Very dark grayish-brown (10 YR 3/2) clay loam, lighter when dry; massive structure; slightly plastic; fine roots; clear smooth boundary.

(B)₁ (40-90 cm): Brown (7.5 YR 5/4) to strong brown (7.5 YR 5/6) clay, with red (2.5 YR 4/8) spots becoming larger with depth; inclusions of light yellowish-brown (2.5 Y 6/4) less weathered rock fragments (micas); massive structure; slightly plastic; shiny coatings; clear wavy boundary.

(B)₂ (90-200 cm): Yellowish-red (5 YR 5/8) clay; massive structure; friable, slightly plastic; gradual boundary with the mottled zone which is less clayey and progressively paler.

M.Z. (Photo XVIII₃): Strongly weathered granitic material, with large red (2.5 YR 4/8), yellow (2.5 Y 8/8), and white (2.5 Y 8/0) mottles; massive structure. NOTE: *These differences in color are produced by varying hydration or even localized water-saturated conditions.*

Geochemical and Biochemical Properties

Particle-size distribution: While the A₁ horizon is strongly "depleted" (28% clay), clay content is essentially constant (56-58%) from 40 to 200 cm. It does not appear that true clay "illuviation" has taken place; clay coatings are weakly expressed and are probably "stress-cutans."

Weathering products: *Kaolinite* predominates in the clay-size fraction, but a small amount of illite is detectable in (B)₁. Illite is partially transformed into vermiculite within the humic A₁ horizon. Additionally, (B)₁ contains about 5% Al_2O_3 and 5% Fe_2O_3, both in the amorphous or cryptocrystalline state.

Exchange complex: As the exchange capacity in (B) is close to 26 meq/100 g of clay, this soil cannot be classified with Oxisols. The profile is desaturated: S/T is 9% in (B)₁ and increases to 32% in (B)₂. The value of S increases from 1.3 meq/100 g at the surface to 4.6 meq/100 g in (B)₂.

Biochemical properties: Medium organic matter content at the surface (3% O.M. with a C/N ratio of 9.2). (B)₁ still contains 1.7% organic matter.

Genesis. This soil is difficult to classify. It is really a *young Ferralitic soil* as indicated by the inclusions of incompletely weathered rock fragments and by the presence of micaceous minerals. At high elevation, ferralitization is slower and, here, it is further impeded by the slope. The humic horizon is thick—a typical feature of soils at high elevation. The acid organic matter accelerates the weathering process by inducing "microfragmentation" and even sheet opening of the micaceous minerals in A₁. At still higher elevations, organic matter is even more abundant and more acid and weathers a portion of the clay minerals which release gibbsite in the A₁ horizon (Ferrisols with secondary gibbsite). In the upper part of this profile, weathering proceeds in the presence of organic matter and, thus, the slow crystallization of iron oxides produces an ocher color. At depth, the red color is produced by ferrihydrite and hematite which crystallize more rapidly in the absence of organic matter (Schwertmann et al., 1974).

REFERENCE: Faivre, P. and H. Ruisz, Codazzi Institute, Bogota, Colombia, 1973.

XVIII₄₋₆: FERRALITIC HARDPANS

XVIII₄: Ferralitic soil with petroplinthite

Rapidly forming hardpan resulting from absolute accumulation of iron by hardening of a plinthite layer.

Location and ecology: Located in the Llanos Plain, Colombia, a few kilometers from Profile XVII₃, but at a lower elevation (edge of valley), in a ground water seepage zone.

Profile description and properties: Towards the 100-cm depth, there is a discontinuity in parent materials: an old alluvial deposit underlies a more recent deposit. The hardpan has developed at this textural discontinuity, under the influence of laterally flowing ground water. The cemented horizon consists of discontinuous concretions beginning at a depth of about 50 cm and of a 15-cm thick, massive and continuous hardpan from 85 to 100 cm. Properties are similar to those of Profile XVII₃. The only difference is that the former plinthite layer is hydromorphic and has been subsequently cemented.

Genesis (Van Wambeke, 1971). The hardpan has formed in two stages: (i) formation below the water table (hydromorphic conditions of *soft* plinthite at the foot of the slope, along the thalweg; and (ii) overall decline of the water table by deepening of the valley (fluvial erosion). Cementation takes place by desiccation and hardening of the plinthite.

REFERENCE: Faivre, P., Codazzi Institute, Bogota, Colombia. (Photo by P. Faivre.)

XVIII₅: Scoriaceous ferralitic hardpan formed on a plateau

Old, slowly forming hardpan resulting from mixed accumulation.

Location and ecology: Adamaoua table-land, The Cameroons. Elevation 1,120 m. Humid tropical climate. P. 1,575 mm; M.T. 22°C. Vegetation is a secondary savanna resulting from forest clearing. Parent material is acid granite. Topography: high platform.

Profile properties: Very thick hardpan (several meters), partially dismantled by erosion; *scoriaceous* aspect; relatively soft areas, containing sand and kaolinite, separated by a very hard ferruginous matrix, forming a solid and coherent "skeleton" (this contrasts with a *pisolithic* aspect which has hard concretions cemented by a soft ferruginous mass).

Genesis. Former soil-building processes, under forest, led to an initial concentration of sesquioxides by relative accumulation. Then, the absolute accumulation of iron was promoted by the acidity and the relative hydromorphic conditions of the profile. This iron filled cracks and formed the profile matrix. Bush fires and the continual clearing of the forest for several millenia have contributed to accelerate the process and promoted hardening by partial dehydration and crystallization of the ferruginous cements. Erosion has lacerated this compact mass into dislocated platforms.

REFERENCE: Hervieu, 1970. (Photo by S. Bruckert.)

XVIII₆: Hardpan resulting from relative accumulation

Location and ecology: Cape Bocage, New Caledonia on ultra-basic parent material (peridotite). Equatorial climate. P. 2,000 mm; M.T. 23.5°C. Vegetation consists of degraded shrub forest.

Profile properties: Very dark brown, rounded and discontinuous formations of iron oxides, enriched with various trace elements (chromium and nickel oxides, etc.), resting on a weathered and spongy material high in amorphous material. Very little alumina.

Genesis. Typical relative accumulation within a material that was initially poor in silica; total elimination of silica and exchangeable bases; hardening has intensified due to the anthropic degradation of the forest.

REFERENCE: Allouc, J., unpublished report from the Geology Institute, University of Nancy I. (Photo by J. Allouc.)

PLATE XVIII
Ferrisols and Ferralitic Hardpans

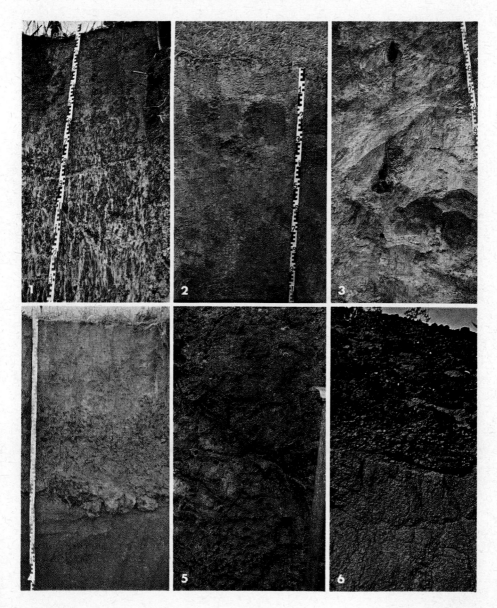

CHAPTER X

SODIC SOILS

Soils whose genesis is influenced by the sodium ion, present as either a salt (NaCl) *in solution (saline soils), or as an* exchangeable *cation on the exchange complex (soils with exchange complex rich in sodium), or both forms at the same time.*

There are two possible origins for the exchangeable Na^+ bound to the complex. It may result from (i) the exchange with Na^+ ions contained in saline ground water (or from a fossil saline deposit) or (ii) from the direct partial saturation of the exchange complex with Na^+ released by the weathering of sodium-bearing minerals. *In both cases, it is necessary that the concentration of* Na^+ *in the soil solution greatly exceed the concentration of divalent cations* (Ca^{2+} *and* Mg^{2+}), *in order that the exchange complex retain a high proportion of* Na^+ *ions.*

Whatever the form it takes, the Na^+ ion is very mobile and will remain in the profile only under *dry* climatic conditions, when high potential evapotranspiration prevents any "climatic drainage." Sodic soils will therefore be found at *site climaxes* (with a typical halophytic vegetation), in steppe, semidesert, or dry tropical regions. They can also be found in coastal areas under more humid climate: e.g., Polder or Mangrove soils. In such cases, the source of sodium is either seawater or lagoon water with a composition close to that of seawater. In these areas, Sodic soils are always characterized by the presence of saline ground water.

DEFINITIONS AND PRINCIPLES OF CLASSIFICATION

The relative proportion of divalent cations (Ca^{2+} and Mg^{2+}) on the exchange complex with respect to Na^+ ions is of great importance in classifying Sodic soils.

Calcic Solontchaks. Saline soils with exchange complex rich in calcium (Na^+ represents less than 10-15% of exchangeable cations). The profile is weakly differentiated, but soil structure is strongly expressed; generally, a shallow saline water table is present. Artificial desalinization does not lead to alkalinization of the profile.

Sodic Solontchaks. Saline soils with exchange complex rich in sodium (Servant, 1970). Also characterized by a shallow saline water table, but contrary to the previous soil, Na^+ represents a nonnegligible proportion of exchangeable cations (up to 30%). These soils are typical of coastal lagoons and constitute the initial stage of development of "Polder soils," before the land was diked.

Alkali soils. Soils with a strong alkaline reaction due to the presence of Na^+ ions on the exchange complex. (Na^+ represents 30-50% of exchangeable cations; pH exceeds 8.5.) pH values are all the more alkaline as exchangeable Na^+ is more abundant and as the concentration of NaCl in the soil solution is low. Soil structure is destroyed as sodium-clays induce a strong plasticity and anaerobic conditions in the profile. The organic matter which is dissolved under alkaline conditions frequently forms a superficial black crust (black alkali). Depending upon the origin of the sodium, two types of alkali soils will be defined:

Weakly saline alkali soils. Na^+ comes from NaCl in the ground water. These soils derive from Sodic Solontchaks after moderate desalinization.

Nonsaline alkali soils. Na^+ usually comes from the weathering of sodium-bearing minerals under dry climate.

Solonetz. Soils that are characterized by the translocation of sodium-clays. A *natric horizon* develops in which the structure is prismatic at first [desiccation cracks in the (B) horizon of an alkali soil], and thereafter becomes *columnar*. Peds are often coated with black, amorphous material containing organic matter.

Soloths. Soils that result from the superficial acidification of Solonetz. Soloths have an acid humus and a bleached A_2 horizon. This horizon is loamy in texture, high in silica and contains small concretions produced by impermeabilization of the surface and iron segregation during temporary waterlogged conditions. The former natric horizon can sometimes still be present at depth, but its structure is more or less rapidly degraded.

GENESIS OF SODIC SOILS

Two fundamental genetic processes may be identified according to the two possible sources of sodium.

Seasonal or permanent desalinization

Desalinization may result either from a seasonal dilution of the soil solution or from a decline of the saline water table, occurring naturally or artificially (irrigation, drainage, diking of polders). Fresh water, in the form of rainwater or irrigation water, has an increasing influence on soil genesis.

I will not discuss at length the case of *calcium-rich ground water* (Calcic Solontchak). The decline of the water table does not generally induce important modifications in the soil properties, precisely because of the predominance of divalent cations (mainly Ca^{2+}) on the exchange complex. In many profiles, however, some of the calcium precipitates as gypsum (gypsic soils; Dutil, 1971).

Genesis in the presence of ground water in which NaCl predominates is much more complicated because of the partial saturation of the exchange complex with Na^+. The weak affinity of the exchange complex for monovalent cations (versus divalent cations) is compensated by the much greater concentration of sodium in the ground water. Therefore, Na^+ represents a nonnegligible proportion (up to 50%) of exchangeable cations.

Sodium-clays tend to hydrolize in the presence of fresh water and induce the alkalinization of the profile (pH exceeds 8.5 and may reach 9 or even 10) and the subsequent degradation of soil structure. However, alkalinization is slowed in the presence of brackish water, as NaCl lowers the pH and maintains the clays in the flocculated state (aggregated structure).

When the saline water table is lowered, desalinization is accompanied by large changes not only in pH, but sometimes also in Eh, which induces oxidation if the medium was initially reducing. Three cases will be considered:

Nonreducing initial medium with low sulfide content. The dominant process is alkalinization by partial hydrolysis of sodium-clays in a slightly saline medium (slightly saline alkali soil). Alkalinization is often accompanied by translocation of sodium-clays and formation of a natric horizon (Solonetz, which may in turn evolve into Soloths through superficial acidification).

Some Solonetz form by evolution of alkali soils following the decline of the saline water table. Examples are the steppe Solonetz in the southeastern U.S.S.R. Others result from an important seasonal change in the salinity of the soil solution. During the dry season, the exchange complex is enriched with sodium; during the wet season, sodium-clays are dispersed and eluviated (Humic Solonetz with saline ground water) (Kovda, 1965; Szabolcs, 1969).

Reducing initial medium with presence of iron sulfide. The saline soil with exchange complex rich in sodium is initially characterized by the presence of black iron sulfide spots, which are indicative of a very low Eh (Polder soils, Mangrove soils with sulfate reduction) and evolves as follows:

1. *In a non- (or slightly) calcareous medium*, the major process is *acidification through oxidation of sulfides*. Iron is oxidized and precipitates as mottles. Simultaneously, the formation of sulfuric acid in a non- or slightly buffered environment sometimes leads to extreme acidification.

2. *In a calcareous medium*, the presence of sufficiently large amounts of$CaCO_3$ neutralizes acidification by formation of calcium sulfate. This can be observed in the "Calcareous" or "Calcic" Polder soils of Northern Germany. But this does not exclude a subsequent evolution of such Polder soils. Under humid climate, carbonates are leached out, followed by migration of clay minerals which generates a plugged B_t horizon. Secondary hydromorphic and acid conditions then develop very gradually with formation of peat (Peaty Polder soils).

Weathering of rock containing sodium-bearing minerals

When parent rock contains sodium-bearing minerals (albite, nepheline, oligoclase), the amount of Na^+ ions released by weathering is often large enough to induce the *direct alkalinization* of the exchange complex (with no contribution from saline ground water) provided that the climate is dry enough to preclude any loss of sodium.

In order that alkalinization may take place, it is necessary that, during some periods of the year, the Na^+ concentration in the soil solution largely exceed the Ca^{2+} concentration and, to some degree, even that of Mg^{2+}. This is made possible by the different solubility products of the salts of these cations.

During the rainy season, the profile is often temporarily waterlogged and active weathering releases Ca^{2+}, Mg^{2+}, and Na^+. Upon drying, the less soluble cations precipitate, first Ca^{2+}, then Mg^{2+}. The exchange complex will preferentially adsorb those cations that remain in solution, mainly Na^+ and accessorily Mg^{2+} (Kovda, 1973).

Similarly, when ground water flows down the slopes of a tropical "Inselberg", it becomes gradually depleted of Ca^{2+}, then of Mg^{2+}, and finally contains only Na^+ ions. At the foot of the slope, Solonetz or Soloths will be found (Bocquier, 1973).

APPENDIX: ARID CLIMATE SOILS

Although Sodic soils are well represented, they are not the only ones that characterize arid climatic zones. Sierozems (subdesert Gray soils) are also found and their place is poorly defined in the various soil taxonomies. Whereas the U.S. and F.A.O. systems of soil classification maintain a special order (Aridisols), in the French system, they are included with the "Immature soils." Other taxonomies consider them as soils of the driest steppes, thus low in organic matter. Nevertheless, some of their properties are similar to those of Sodic soils in dry regions (for example, a relatively high content of exchangeable Na^+ on the exchange complex). Therefore, I have included one Sierozen (Profile XIX_6) in this chapter for purposes of comparison.

Table 12.   Genesis of Sodic Soils[a]

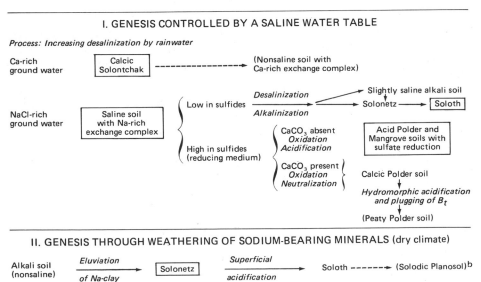

I. GENESIS CONTROLLED BY A SALINE WATER TABLE

Process: Increasing desalinization by rainwater

Ca-rich ground water → Calcic Solontchak ------------→ (Nonsaline soil with Ca-rich exchange complex)

NaCl-rich ground water → Saline soil with Na-rich exchange complex →
- Low in sulfides → *Desalinization* / *Alkalinization* → Slightly saline alkali soil ↓ Solonetz → Soloth
- High in sulfides (reducing medium) →
  - $CaCO_3$ absent *Oxidation Acidification* → Acid Polder and Mangrove soils with sulfate reduction
  - $CaCO_3$ present *Oxidation Neutralization* → Calcic Polder soil ↓ *Hydromorphic acidification and plugging of B_t* ↓ (Peaty Polder soil)

II. GENESIS THROUGH WEATHERING OF SODIUM-BEARING MINERALS (dry climate)

Alkali soil (nonsaline) → *Eluviation of Na-clay* → Solonetz → *Superficial acidification* → Soloth ------→ (Solodic Planosol)[b]

[a]Sources: Kovda (1965), Brummer et al. (1971), Cheverry (1974), and Vieillefon (1974).
[b]See Profile XV_5 (Chapter VII).

XIX₁: CALCI-MAGNESIAN SOLONTCHAK

(Solontchak calci-magnésique)

F.A.O.: *Orthic Solontchak*; U.S.: *Typic Gypsiorthid*

Location: Giguela Valley, Ciudad Real, Spain.
Topography: Level alluvial plain.
Parent material: Recent alluvium enriched with gypsum from slopes with Tertiary outcrops.
Climate: Semi-arid Mediterranean with dry summer season. P. 400 mm; M.T. 14.5°C.
Vegetation: Halophytes consisting of *Suaeda vera, Frankenia reuteri.*

Profile Description

A_p (0-22 cm): Dark brown to brown (10 YR 4/3) humic horizon; crumb structure, aerated; common roots; white evaporite specks during the dry season; clear boundary.

A(B) (22-70 cm): Very pale brown (10 YR 7/3) gypsic horizon; massive structure, compact; white gypsum evaporite specks (can be seen in the old cut on the right-hand side of the photograph); effervescent; few roots; clear boundary.

C: Reddish alluvium; loose.

Geochemical and Biochemical Properties

Overall composition: The silicate fraction only represents 30% (nearly half of which is clay) of the total soil mass in A_p and 5% in A(B). This horizon contains approximately 89% gypsum and 5% $CaCO_3$.

Salinity and exchange complex: Electrical conductivity of the saturated paste extract is 55 mmhos/cm in A_1 and 31 mmhos/cm in the gypsic A(B) horizon. Thus, this is a saline soil. The Na^+ ion (as NaCl) is a minor cation in the saline ground water; it represents 20 to 30% of total cations. Under these conditions, exchangeable Na^+ represents only 8-9% of total exchangeable cations at 30 cm and 4% at 80 cm. As Mg^{2+} is more soluble than Ca^{2+}, which precipitates rapidly as $CaSO_4$ and $CaCO_3$ during the dry periods, it is preferentially adsorbed and predominates over Ca^{2+} on the exchange complex. Under these conditions, pH remains high [8.5 in A_p, 8.8 in A(B)].

Biochemistry: The organic matter (2.6% in A_p with a C/N ratio of 13) is flocculated into aggregates by the divalent cations.

Genesis. Because of the large reserves of gypsum and dolomite which exist in the uphill Tertiary deposits, the saline ground water contains more divalent cations (calcium and magnesium) than monovalent cations (sodium). Under these conditions, the exchange complex will preferentially adsorb divalent cations, while Na^+ represents much less than 15% of total exchangeable bases. Thus, this soil is a "Calcic Solontchak." But it has a special feature compared to typical Calcic Solontchaks—Mg^{2+} predominates over Ca^{2+} on the exchange complex. This explains the abnormally high pH value of the soil. For this reason, I think that a better designation for this soil is *Calci-magnesian Solontchak.* When such a soil undergoes desalinization by a decline of the saline water table and by irrigation with fresh water, *alkalinization* does not occur as it does in saline soils with a Na-rich exchange complex (Sodic Solontchak).

REFERENCE: Porta, 1973.

XIX₂: SALINE SOIL WITH SODIUM-RICH EXCHANGE COMPLEX
(drained)

(Sol salin à complexe sodique)

F.A.O.: *Orthic Solontchak*; U.S.: *Typic Halaquept*

Location: Guadalquivir marshland, Spain.
Topography: Alluvial plain. Elevation 3 m.
Parent material: Recent alluvium of the Guadalquivir River.
Climate: Warm Mediterranean, with dry season. P. 500 mm; M.T. 19.5°C (at Sevilla).
Vegetation: Crops on drained land; native vegetation consists of halophytes (*Obione, Salicornia*).

Profile Description

A_1 (0-10 cm): Gray (10 YR 5/1) slightly humic loam; crumb structure; many roots.

A_2 (10-30 cm): Light gray to gray (10 YR 6/1) clay loam to clay; single-grained to massive structure; firm; beginning of blocky structure formation by irregular cracking; gradual boundary.

B_t (30-60 cm): Light brownish-gray (10 YR 6/2) clay; prismatic structure with vertical cracks, no coatings; gradual boundary.

(B)C: Brown (7.5 YR 5/2) clay; tendency towards cubic structure; gradual transition to massive fine alluvium.

NOTE: This profile is drained and cultivated; the water table has been lowered by about 60-70 cm with respect to a neighboring nondrained profile.

Geochemical and Biochemical Properties

Particle-size distribution: Fine alluvium with a loam to clay texture. A beginning of clay eluviation can be observed. Clay content is 25% in A_1, 40% in A_2, 52% in B_t, and slightly less in (B)C. However, no clay coatings can be detected in B_t.

Salinity and exchange complex: Ground water is saline and contains *essentially sodium*. Electrical conductivity is 20 mmhos/cm below 60 cm, but very low at the surface (1-2.4 mmhos/cm) due to the lowering of the water table. Below 60 cm, SAR (sodium adsorption ratio) of the ground water exceeds 30, which indicates a very high predominance of Na^+ (as NaCl) over Ca^{2+}. The ratio of exchangeable sodium to total exchangeable cations (Na^+/S) is low at the surface (0.04 in A_1), but increases in A_2 and B_t to about 0.20 (pH in water 8.6) and to 0.31 in (B)C, while pH remains below 8 because of the strong salinity. The $CaCO_3$ content is approximately 15% at the surface and 20-30% at depth.

NOTE: A neighboring *nondrained* profile displays features that are more homogeneous: conductivity is 28 mmhos/cm *at the surface*; the Na^+/S ratio exceeds 0.3 in A_1 and A_2, but the pH of the soil solution remains below 8.

Organic matter: Organic matter content is 2.4% at the surface and 0.4-0.5% at depth. Alkalinity is still too low to provoke its mobilization.

Genesis. The comparison between the properties of two neighboring profiles with a saline water table—one a nondrained profile, the other a drained and cultivated profile (presented here)—allows the identification of the initial stages of transformation of a saline soil with sodium-rich exchange complex into an alkali soil only slightly saline at the surface, as influenced by the fall of the saline water table. The nondrained profile is weakly differentiated from both a morphological and a geochemical viewpoint. The influence of the saline ground water is evident throughout. Despite a Na^+/S ratio which reaches or exceeds 0.30, pH remains below 8. Under the action of surface desalinization, the profile becomes differentiated. Although the lower horizons preserve the properties of a "saline" soil (low alkalinization), the surface horizons evolve by dispersion and migration of a small fraction of the sodium-clays. The Na^+/S ratio decreases. However, pH increases somewhat because fresh water (rain) causes the hydrolysis of the remaining fraction of sodium-clays. Evolution has gone beyond the stage of alkali soil with sodium-rich exchange complex. The profile displays some features of an immature Solonetz.

REFERENCE: *Conference on Mediterranean Soils*, Center of Edaphology, Madrid, 1966.

XIX$_3$: SOLONETZ

(Solonetz)

F.A.O.: *Orthic Solonetz*; U.S.: *Typic Natriboralf*

Location: 110 km north of Tselinograd, Kokchetov road, Kazakhstan, U.S.S.R.
Topography: Depression in an undulating plain. Elevation about 300 m.
Parent material: Granite with sodium-bearing minerals (oligoclase).
Climate: Dry and cold continental. P. about 300 mm; M.T. Jan. −18°C, July +19.5°C.
Vegetation: Steppe with *Artemisia* sp. and *Stipa capillata.*

Profile Description

A$_1$ (0-5 cm):   Very dark grayish-brown (10 YR 3/2) humic loam; crumb structure, aerated; gradual boundary.

A$_2$ (5-11 cm):   Grayish-brown (10 YR 5/2) loam; single-grained to massive structure; accumulations of white siliceous powder at the base; abrupt boundary.

B$_t$ (11-30 cm):   Dark brown (7.5 YR 3/2 to 3/4) clay loam to clay, natric horizon, very compact; strong columnar structure, rounded tops covered with a white powder, columns coated with amorphous material rich in organic matter, gradual transition to subangular blocky structure.

C$_{ca}$ (30-50 cm):   Reddish-yellow (5 YR 6/6) clay loam with many white spots of CaCO$_3$ and gypsum; gradual boundary with the C horizon, rich in gypsum crystals.

Geochemical and Biochemical Properties

Particle-size distribution: Abrupt change in texture from the A horizons (10% clay) to the natric horizon (more than 40% clay). Clay content decreases progressively from the base of the B$_t$ to the C horizon (35%). The siliceous powder on top of the columns is the sign of alkaline degradation of the clays.

Exchange complex: Exchange capacity is 30 meq/100 g in the B$_t$ horizon. The exchange complex is saturated with Ca^{2+} (33% of S), Mg^{2+} (48% of S), and Na$^+$ (16% at the top and 30% at the base of B$_t$). The gradual increase in saturation with Na$^+$ corresponds to increasing alkalinity with depth: pH (water) is 7.6 in A$_2$, 8.2 in B$_t$ and 9 or even 9.3 at the base of B$_t$.

Biochemistry: Organic matter content is 6% in A$_1$, 3.5% in B$_t$, and 2.5% at the base of B$_t$. This indicates solubilization and migration of humus due to alkalinity (amorphous organic coatings around the "columns"). "Humin" is the major constituent (82%) of this organic matter and results from the transformation of alkali-soluble substances by aging.

Genesis. In this profile, Na$^+$ does not come from saline ground water but from direct release by weathering of sodium-bearing minerals. The absence of "climatic drainage," associated with high potential evapotranspiration, prevents any leaching of Na$^+$ ions. The genesis of this profile involves two phases, both of which are linked to contrasts in soil climate—short periods of saturation during the rainy season alternating with prolonged desiccation periods. The *first phase* corresponds to "alkalinization" of the profile. During the rainy season, weathering releases Ca^{2+}, Mg^{2+}, and Na$^+$ cations. During dry periods, Ca^{2+} precipitates in the C$_{ca}$ horizon as powdery CaCO$_3$ spots and gypsum crystals, thus promoting the preferential adsorption of Mg^{2+} and especially Na$^+$ on the exchange complex. There is a gradual increase in pH. The *second phase* takes place during the wet seasons. Clay minerals and sodic humates are dispersed and migrate to form the natric horizon, with a simultaneous corrosion of the column faces by transformation of clays into amorphous material due to the strong alkalinity.

REFERENCE: *Field trip 5 to North Kazakhstan, 10th International Congress of Soil Science*, Moscow, 1974. (Photo by Guckert.)

XIX₄: SOLOTH
(Soloth)

F.A.O.: *Solodic Planosol*; U.S.: *Albic Glossic Natraqualf*

Location: Oulianovsk School of Agriculture, U.S.S.R.
Topography: Depression, in a plain dominated by Chernozems. Elevation 130 m.
Parent material: Loam covering a terrace of the Volga River.
Climate: Cold continental. P. about 400 mm; M.T. 3.5°C (Jan. −13°C, July +20.5°C).
Vegetation: Crops on former grass steppe.

Profile Description

A_p (0-25 cm):   Grayish-brown (2.5 Y 5/2) sandy loam; massive to blocky structure; compact, slightly porous; abrupt boundary.

A_2 (25-52 cm):   Light gray (2.5 Y 7/2) sandy loam; platy structure; very compact; fine black concretions; clear boundary.

A/B (52-73 cm):   Loam, large and irregular brown areas contrasting with white (10 YR 8/2) powdery spots; subangular blocky structure, compact; dark brown humic clay coatings; clear boundary.

B_t (73-120 cm):   Brownish-yellow (10 YR 6/6) clay loam; prismatic structure, white powdery spots on pedsurfaces clay-humus coatings; fine concretions.

C: Brown loam; massive structure; slightly effervescent.

NOTE: A section of the profile was freshly cut to show soil structure and the cutans.

Geochemical and Biochemical Properties

Particle-size distribution: A clear textural change exists between A_2 (5% clay) and B_t (29% clay). The illuviated cly accumulates over a great depth. Clay-humus cutans are found on the surfaces of the structural prisms.

Exchange complex: Exchange capacity is low in A_2 (5 meq/100 g) and reaches 24 meq/100 g in B_t. The A_p horizon is almot saturated because of cultivation, but a pH of 5 in A_2 and B_t indicates marked desaturation (S/T 60%). Among exchangeable cations, Mg^{2+} and Na^+ are the signs of a previous phase of alkalinization; Mg^{2+} represents 50% of S and Na^+ represents 5% of S in the nonacidified horizons.

Biochemistry: Organic matter content is 4.2% in A_p with a C/N ratio of 16.8. It is very low in A_2 but increases slightly in A/B and B_t (dark cutans). The organic matter is dominated by fulvic acids (FA/HA ratio is 3.5). Fulvic acids form complexes which are likely to mobilize iron, but the translocation index for iron does not exceed 1/3, because of the presence of concretions in A_2.

Genesis. Soloths constitute the utmost extent of evolution of soils with sodium-rich exchange complex. Evolution starts with the dispersion and translocation of the sodium-clays (Solonetz), followed by the acidification and degradation of the exchange complex by rainwater. In this profile, initial alkalinization was caused by a lowering of the saline water table (with NaCl). The action of rainwater has been emphasized by the special location of this soil in a depression which collects rainwater and snowmelt water. Accordingly, sodium-clays have migrated and have been "de-alkalinized," and even acidified and degraded during seasonal periods of water saturation. Supporting signs of this evolution can be found in the platy structure of the A_2 horizon, the white siliceous spots in A/B and B_t (along the prisms), the segregation of iron and aluminum caused by hydromorphism (concretions), and, finally, the migration and accumulation of fulvic-sesquioxide complexes in the B_t horizon. This profile shows some analogy with Solodic Planosols (Profile XV₅).

REFERENCE: *Guidebook to the Volga-Don excursion, 10th International Congress of Soil Science*, Moscow, 1974.

XIX$_5$: SALINE SOIL WITH SULFATE REDUCTION (Mangrove soil)

(Sol salin à sulfate-réduction)

F.A.O.: *Thionic Fluvisol*; U.S.: *Typic Sulfaquent*

Location: Casamance, Balingore, Senegal.
Topography: Bank along a branch channel in-the estuary of the Casamance River.
Parent material: Marine-fluviatile alluvium consisting of ferralitic material.
Climate: Subguinean tropical. P. 1,500 mm; M.T. 30°C.
Vegetation: Mangrove with *Avicennia nitida* and halophytic succulent plants.

Profile Description

A$_1$ (0-5 cm):   Dark grayish-brown (10 YR 4/2) clay; crumb to blocky structure; mottles; fine salt crystals.
A$_1$G$_0$ (5-20 cm):   Gray (10 YR 5/1) clay; massive structure, prismatic tendency; mottles.
A$_1$G$_r$ (20-55 cm):   Gray clay; massive structure; brown mottles around roots; wavy boundary.
G$_r$ (55-80 cm):   Dark bluish-gray (5 B 4/1) clay; massive structure; very plastic; abrupt smooth boundary with underlying light gray gley.

Geochemical and Biochemical Properties

Particle-size distribution: Clayey material (75-78% clay) usually saturated with water (72% at the surface, 160% at 80 cm). Clay is dominantly kaolinite. Sand content (quartz) is 2-3%.

Salinity and exchange complex: Electrical conductivity varies according to the season from 20 to 70 mmhos/cm, which allows this soil to be classified with "saline soils." Ground water is characterized by a Cl^-/SO_4^{2-} ratio between 2 and 3 and a strong predominance of Na^+ over divalent cations. In particular, very little Ca^{2+} is present. The composition of the exchange complex cannot be precisely determined, but Na^+ is definitely dominant.

Sulfur cycle: The sulfur cycle plays a fundamental role in the genesis of this soil. Sulfur is present in two forms, as *reduced* sulfur (sulfide: pyrite and an "insoluble residue," which is a poorly crystallized form of reduced sulfur, according to Vieillefon, 1974) and as *oxidized* sulfur (soluble sulfate, but no insoluble "jarosite"). Total sulfur content increases greatly with depth. Sulfide-sulfur ranges from 0.3 to 3% at 60 cm. Sulfates amount to 0.14% at depth; they prevent any alkalinization of the horizon and reduce the pH to *2.1* at this level.

Biochemistry: As in many alluvial soils, the distribution of organic matter with depth is irregular (8.4% O.M. at the surface, then 3.6%, and again 8.1% at the 50-cm depth).

Genesis. Mangrove soils are saline Alluvial soils, rich in organic matter and *sulfides* during their initial phase of development. The permanently shallow water table ensures constant anaerobiosis. The profile later develops into a *tanne** by seasonal lowering of the water table, superficial aeration, and more or less complete oxidation of sulfides (Vieillefon, 1974). The sulfuric acid which forms prevents any alkalinization and may even cause the profile to become extremely acid. This profile has reached an intermediate stage of development. The saline water table falls periodically to about the 50-cm depth. Sulfide oxidation is still moderate and does not lead to the formation of insoluble sulfate (*jarosite*) as occurs in more aerated *tanne*. However, enough soluble SO_4^{2-} anions are formed and leached down to acidify the deeper horizons.

REFERENCE: Vieillefon, 1974. (Photo by J. Vieillefon.)

*Translators' note: *Tanne* is an African word meaning partially dried saline soil unsuitable for cultivation.

XIX$_6$: SIEROZEM

(Sierozem)

F.A.O.: *Calcic Xerosol*; U.S.: *Typic Calciorthid*

Location: Kirov State Farm, 40 km southeast of Tachkent, U.S.S.R.
Topography: Piedmont zone, gentle slope towards the southwest.
Parent material: Middle Quaternary calcareous loess.
Climate: Arid. P. 359 mm; M.T. Jan. $-1.1°C$, July $+26.7°C$.
Vegetation: Sparse xerophilous steppe with *Poa bulbosa, Carex pachystylis, Psoralea drupacea, Aegilops cylindrica*, etc.

Profile Description

A$_1$ (0-5 cm):   Grayish-brown (10 YR 5/2) silt loam; fine crumb structure; aerated; common roots; clear boundary.

A/C (5-35 cm):   Pale brown (10 YR 6/3) silt loam; coarse crumb structure; compact; animal burrows; small carbonate concretions at the base; gradual boundary.

C$_{ca}$ (35-70 cm):   Very pale brown (10 YR 7/3) silt loam; coarse subangular blocky structure; few roots; much pseudomycelium and calcareous nodules; gradual boundary.

C (70-115 cm):   Paler silt loam; massive to subangular blocky structure; few carbonates.

Geochemical and Biochemical Properties

Particle-size distribution: Texture is typical of loess with a strong predominance of the silt fraction and very little sand. Clay content is minimum in the C horizon (12-13%), it increases towards the surface to reach a maximum around the 20- to 30-cm depth (18-20%). Inherited illites are the dominant clay minerals, but paticle-size distribution suggests that some clays were formed by weathering in the middle horizons.

Exchange complex: The exchange capacity is moderate (13 meq/100 g in A$_1$, 10 meq/100 g in A/C). The exchange complex is saturated, mainly with Ca^{2+} but also with some Mg^{2+} and traces of Na^+. Carbonates are found throughout the profile (6% in A$_1$ and a maximum of 20% in C$_{ca}$). pH (water) varies from 7.4 to 7.5.

Biochemistry: For a Sierozem, this soil is relatively rich in organic matter (3.7% in A$_1$ and still 1.2% at the top of A/C). The C/N ratio is very low (8.5 in A$_1$), which is a constant feature. Due to the decomposition of deep grass roots, the mineral horizons contain 0.5% organic matter. The FA/HA ratio is close to 1 in A/C, which is a higher value than that found in Isohumic soils.

Genesis. Although precipitation is not negligible, the soil climate of this profile is drier than that of the driest Isohumic soils, due to a higher potential evapotranspiration rate and also to the extreme irregularity of rainfall. The organic matter content is quite high in A$_1$, but decreases more rapidly with depth than in Isohumic soils. Furthermore, the mineral matter is not very weathered; the pale brown color of the profile is indicative of the low amounts of free iron being released. Nevertheless, weathering is not entirely absent judging by the maximum clay content observed near the 20- to 30-cm depth. This clay probably originates from the comminution of the coarser micaceous minerals contained in the parent material. Although the surface horizons are not decarbonated, carbonates do migrate to some extent and form a C$_{ca}$ horizon. Thus, certain common features with Isohumic soils are present here and it may be concluded that this soil follows Isohumic soils in a climatic sequence from semi-arid to arid soils.

REFERENCE: *Guidebook to field trip 4, 10th International Congress of Soil Science*, Moscow, 1974. (Photo by Maucorps.)

PLATE XIX

Sodic Soils and Arid Soils

SOIL STRUCTURE AND MICROSTRUCTURE

I have frequently used criteria of soil structure and microstructure as diagnostic elements of the genesis of the profiles I have described. It would seem appropriate to give some explanations concerning the interpretation of these data by taking examples from the profiles that were studied.

Without going into the details of the terminology used (which I feel is often unnecessarily complicated), I will simply give a few essential definitions concerning soil structure and microstructure.

DEFINITIONS

1. Plasma. The plasma includes all fine particles likely to form the cement between the coarser elements of the *skeleton*. The plasma may be flocculated into distinct aggregates (*agglomerated* microstructure) or, on the contrary, it may be dispersed and form a compact mass (*massive* microstructure), often called "lehmified" microstructure by Kubiena (1953).

2. Cutans. Cutans are zones where the plasma is superficially differentiated. They are located on the surfaces of peds or along channels (or pedotubules). There are two kinds of cutans:

(i) Stress-cutans are friction surfaces which are caused by structural units sliding past each other. These movements produce *polished* surfaces, which are often shiny. The diagonally oriented and often grooved *slickensides* of Vertic soils are typical examples.

(ii) Illuviation cutans (or *coatings*) result from a deposition in thin films on the surfaces of peds or on the coarse elements of the skeleton. Illuviation cutans are a sign of *migration* in the soil, contrary to stress-cutans which are formed *in situ*.

One can distinguish between *clay skins*, which consist of stacked crystalline clay particles (argillans often colored with iron oxides, or ferriargillans) and are easily recognizable under the polarizing microscope, and *amorphous coatings* which are either organic (gray or black), mineral (ocher), or of mixed composition.

3. Inclusions. Inclusions vary widely in shape, aspect, and composition. *Glaebules* are rounded inclusions, often with a concentric structure;

papules have sharp external boundaries and are often formed of disrupted argillans; "floccules" are silt-size spherical porous aggregates, some are of coprogenic origin, others result from chemical precipitation.

USE OF STRUCTURE AND MICROSTRUCTURE IN THE STUDY OF PROFILE GENESIS

Only three examples will be given here. They concern (i) the study of humus, (ii) the study of mineral diagnostic horizons, and (iii) the study of "complex" or "sequential" profile developments (polycyclic soils).

Study of humus types

The microstructure of A_1 horizons of mulls contrasts with that of moders (or mors). Mulls are characterized by built "aggregates," which are cemented by organomineral substances and include some elements of the skeleton. Moders contain juxtaposed skeletal elements (clear or "bare" quartz grains), rounded humic coprogenous aggregates, and "identifiable" organic elements (plant debris).

Study of mineral or mixed diagnostic horizons

Vertic horizons: Presence of desiccation cracks during the dry season and of *slickensides.*

(B) horizons of alteration (cambic): Blocky or prismatic structure without coatings.

Argillic-type B_t horizons: Formation by illuviation of clay skins. *"Primary" illuviation clay skins* (Fedoroff, 1966; De Coninck and Herbillon, 1969) result from the simultaneous migration of clay and ferric iron; coatings are colored with iron—*ferri-argillans.* *"Secondary" illuviation clay skins* result from eluviation under acid and poorly aerated conditions which leads to the more or less complete segregation of clay and iron (see Profile X_1)—*bleached or white argillans.*

Spodic-type horizons: Accumulation of amorphous material resulting from podzolization. These horizons display either (i) a structure with amorphous "films" cementing quartz grains together more or less efficiently (*alios*), or (ii) a structure with spherical "intergranular aggregates" (pellets), which is characteristic of some humic spodic horizons. This microstructure may be of chemical origin by fragmentation of the films (Altemüller, 1962; Franzmeier et al., 1963) or of biological origin by formation of coprogenous elements following the digestion of precipitated organic matter by lower organisms (De Coninck et al., 1974).

The overall result is a *fluffy structure* which is characteristic of some spodic horizons or horizons with a spodic trend (Profiles IX_3 and XI_1, for example).

Hydromorphic horizons: Gley or pseudogley always characterized by the segregation of iron which concentrates in *mottles* or *concretions.* *Oxidized gleys* exhibit amorphous or paracrystalline iron oxide films on ped surfaces (Blume, 1968). *Pseudogleys* are characterized by the presence of glaebules or diffuse mottles within structural units.

Study of multiple or complex pedogeneses

In profiles characterized by polycyclic or complex genesis, the illuvial horizons often display several generations of coatings (or aggregates) whose nature and structure are different. They result from successive illuviation phases and are indicative of the different stages of profile development. The oldest coatings are frequently mechanically disrupted (by cryoturbation, for example) and are then called "papules" which constitute a particularly important diagnostic criterion.

Secondary Podzols resulting from "indirect" podzolization, i.e., subsequent to an eluviation phase. Generally, argillic- or spodic-type horizons are superposed, but intermediate stages exist in which both types of microstructure coexist because the two processes occur simultaneously (B_tB_s horizon, Profile XII_2).

Soils with glossic horizon or with fragipan. Old glossic and fragipan structures can be detected by micromorphological analysis. Old ferri-argillans, usually very thick, are modified in the following manner: in the upper part of B_t horizons, they are mechanically disturbed (by cryoturbation) and form irregular "papules"; in the deep desiccation cracks of fragipans, they remain undisturbed but they have been deferrified and degraded by a reduction in the degree of crystallinity of clay minerals (*degradation argillans*). This can be seen in Profiles X_3 and X_4; Profile XX_6 shows a macroscopic view of a "tongue."

Profiles of complex Eluviated soils. Profile IX_6 is a characteristic example from which Micrographs XX_8 and XX_9 were taken. The B_t horizon is of mixed origin (terra fusca mixed with loam), as is shown by disrupted papules of rubefied terra fusca located at the base of the horizon, while the top of the horizon has been subsequently enriched with illuviated clay, as indicated by the presence of undisturbed ferri-argillans on the surfaces of peds.

Plate XX: SOIL STRUCTURE AND MICROSTRUCTURE

Captions and comments

XX₁: Crumb structure of a chernozemic humic A₁ horizon and of the underlying calcic A₁ca horizon with pseudomycelium

Danubian Chernozem (Semloc, Tisa River Plain, Romania): The photograph shows the transition from the base of the A₁ horizon with irregular crumb structure ("agglomerated" microstructure) to the top of the A₁ca horizon with precipitated carbonates as "pseudomycelium" disseminated into irregular white spots.

XX₂: Argillic horizon with cubic structure and krotovinas

Eluviated Danubian Chernozem (Profile VII₅, Chapter IV): Under very contrasting continental climate, structural blocks change into cubes. Ferri-argillans form shiny coatings and the presence of *krotovinas* (animal burrows filled with humus) indicate that this is indeed a Chernozem.

XX₃: Detail of a slickenside in a Vertisol

"Smonitza" Vertisol (Cheglevic, Tisa River Plain, Romania): Very large, shiny surfaces are polished and have a diagonal orientation.

XX₄: Microstructure of a spodic B horizon

Tropical Hydromorphic Podzol (Profile XII₆, Chapter VI) (Turenne, J.F., 1975. Unpublished Doctoral Thesis, University of Nancy I, 180 pp.): This type of microstructure clearly illustrates two aspects of pedogenesis: (i) a brown "fluffy" microstructure, forming loose coatings or intergranular aggregates and consisting essentially of amorphous mineral particles; (ii) black films, coating the inside of pedotubules or the faces of cracks and consisting mainly of amorphous organic elements.

XX₅: Fragipan horizon

Horizontal cut (Tizli, Italy): The overall aspect is massive with an ocher plasma and a polygonal network of vertically oriented, bleached cracks. These cracks have been subsequently invaded by roots which have left black traces following their decomposition.

XX₆: Old tongue

Partially rubefied loam, mixed with gravelly terrace material (Sainte-Hélène Forest, Rambervillers, Vosges, France) (Photo by B. Souchier): Successive layers can clearly be observed from the center to the outside of the tongue. (i) Towards the top, the center of the "funnel" is filled with white and very fine quartz powder. (ii) Next, and from the top to the bottom of the tongue, is a layer of grayish or bleached degradation argillans. (iii) Finally, a layer of old, reddish-ocher ferri-argillans appears on the outside of the tongue. The latter are more or less blended with the plasma, but, macroscopically, they can be identified by their vertical orientation. This tongue is connected to a "fragipan"-type network, filled with degradation argillans, with some horizontal ramifications (visible in the photograph).

XX₇: Argillic horizon of a complex Eluviated Brown soil

B_t horizon formed on complex parent material consisting of loam and terra fusca (similar to Profile IX₆, Chapter V): The blocky structure is characteristic of the Atlantic temperate climate and is less pronounced and more irregular than that of argillic horizons under continental climate (Micrograph XX₂). The shiny faces of ferri-argillans are clearly visible. Ferri-argillans are evidence of a pedogenic evolution by illuviation which has been superimposed on the original heterogeneity of the parent material. This is confirmed by examination of the microstructure (Micrographs XX₈ and XX₉).

XX₈: Undisturbed ferri-argillans in a complex argillic horizon

Microstructure in the upper part of the B_t horizon of Profile IX₆, Chapter V (Hetier et al., 1972): Undisturbed ocher ferri-argillans found along desiccation cracks are signs of recent illuviation of material originating in the overlying horizon dominated by silt-size particles.

XX₉: Mechanically disrupted inclusions of terra fusca

Complex argillic horizon (Profile IX₆, Chapter V): Inclusions or "papules" can be seen *at the base* of the B_t horizon of Profile IX₆ (Hetier et al., 1972). These inclusions have been disrupted and disturbed by cryoturbation. They consist of old ferri-argillans as demonstrated by their platy aspect. These ferri-argillans are certainly older than the previous ones (Micrograph XX₈) since they are disturbed. Moreover, their reddish color indicates that they consist of terra fusca. These two micrographs clearly show the complex formation of this B_t horizon. It has been enriched with clay by two processes: initially, cryoturbation has caused the upward movement of terra fusca from the base of the profile and, later, clay has migrated from the surface loam and has accumulated mainly at the top of the B_t horizon.

PLATE XX

SOIL STRUCTURE AND MICROSTRUCTURE

BIBLIOGRAPHY

LITERATURE CITED

Altemüller, H.J. (1962): Beiträge zur mikromorphologischen Differenzierung von durchschlämmter Parabraunerde, Podsol-Braunerde und Humus-Podsol. *Z. Pflanzenern. Dung. Bodenk.* **98**: 247-258.

Bartelli, L.J., and R.T. Odell (1960): Field studies of a clay-enriched horizon in the lowest part of the solum of some Brunizem and Gray-Brown Podzolic soils in Illinois. *Soil Sci. Soc. Am. Proc.* **24**: 390-395.

Bartoli, C. (1962): Première note sur les associations forestières du massif de la Grande Chartreuse. *Ann. Ec. Natl. Eaux For. Nancy* **19**: 325-383.

Becker, M. (1971): Etude des relations sol-végétation en conditions d'hydromorphie dans une forêt de la plaine lorraine. *Thèse Doctorat d'Etat, Univ. Nancy*, 225 pp.

Blume, H.P. (1968): Zum Mechanismus der Marmorierung und Konkretionsbildung in Stauwasserböden. *Z. Pflanzenern. Bodenk.* **119**: 124-134.

Bocquier, G. (1973): Genèse et évolution de deux toposéquences de sols tropicaux du Tchad. Interprétation biogéodynamique. *Thèse Doctorat d'Etat, Univ. Strasbourg*, 364 pp.

Bonneau, M., P. Duchaufour, G. Millot, and H. Paquet (1965): Note sur certains sols vertisoliques formés en climat tempéré. *Bull. Serv. Carte Géol. Als. Lorr.* **17**: 325-334.

Bottner, P. (1971): La pédogenèse sur roches-mères calcaires dans une séquence bioclimatique méditerranéo-alpine du Sud de la France. *Thèse Doctorat d'Etat, Montpellier*, 278 pp.

Boulaine, J. (1957): Etude des sols des plaines du Chélif. *Thèse Doctorat d'Etat, Univ. Alger*, 574 pp.

Bruckert, S., and M. Metche (1972): Dynamique du fer et de l'aluminium en milieu podzolique: Caractérisation des complexes organo-métalliques des horizons spodiques. *Bull. Ec. Natl. Super. Agron. Ind. Aliment.* **14**: 263-275.

Bruckert, S., A. Brethes, and B. Souchier (1975): Humification et distribution des complexes organo-métalliques des sols brunifiés et podzolisés. *C. R. Acad. Sci. (Paris) Ser. D* **280**: 1237-1240.

Brümmer, G., H.S. Grundwaldt, and D. Schröder (1971): Beiträge zur Genese und Klassifizierung der Marschen: III. Gehalte, Oxydationssterfen und Bindungsformen des Schwefels in Koogmarschen. *Z. Pflanzenern. Bodenk.* **129**: 92-108.

Bullock, P., M.H. Milford, and M.G. Cline (1974): Degradation of argillic horizons in Udalf soils of New York State. *Soil Sci. Soc. Am. Proc.* **38**: 621-628.

169

Carballas, T., P. Duchaufour, and F. Jacquin (1967): Evolution de la matière organique des Ranker. *Bull. Ec. Natl. Super. Agron. Nancy* **9:** 20-28.

Cheverry, C. (1974): Contribution à l'étude pédologique des polders du lac Tchad. Dynamique des sels en milieu continental subaride dans des sédiments argileux et organiques. *Thèse, Univ. Strasbourg*, 275 pp.

Chirita, C., C. Paunescu, and D. Teaci (1967): *Solurile Romaniei*. Bucharest, 185 pp.

De Coninck, F., and A. Herbillon (1969): Evolution minéralogique et chimique des fractions argileuses dans des Alfisols et des Spodosols de la Campine (Belgique). *Pedologie* **19:** 159-272.

De Coninck, F., D. Righi, J. Maucorps, and A.M. Robin (1974): Origin and micromorphological nomenclature of organic matter in sandy Spodosols. In G.K. Rutherford, Ed., *Soil Microscopy*. Kingston, Ontario, Canada: Limestone Press, pp. 263-280.

Desaunettes, J.R. (1970): *Livret des sols forestiers de Grèce*. Rome: F.A.O., 82 pp.

D'Hoore, J. (1955): Essai de classification des zones d'accumulation des sesquioxydes libres sur une base génétique. *Afr. Soils* **3:** 66-81.

Dommergues, Y., and P. Duchaufour (1966). Caractérisations pédologiques et microbiologiques des stations Lorraines R.C.P. 40. *Rev. Ecol. Biol. Sol* **3:** 533-547.

Duchaufour, P. (1970): *Précis de Pédologie* (3rd ed). Paris: Masson, 481 pp.

Duchaufour, P. (1973): Action des cations sur les processus d'humification. *Sci. Sol* (3): 151-161.

Duchaufour, P., and F. Jacquin (1966): Nouvelles recherches sur l'extraction et le fractionnement des composés humiques. *Bull. Ec. Natl. Super. Agron. Nancy* **8:** 1-24.

Ducloux, J. (1970): L'horizon Bêta des sols lessivés sur substratum calcaire de la plaine poitevine. *Bull. Assoc. Fr. Etude Sol* (3): 15-25.

Dutil, P. (1971): Contribution à l'étude des sols et des paléosols du Sahara. *Thèse Doctorat d'Etat, Univ. Strasbourg*, 346 pp.

Duvigneaud, P. (1974): *La Synthèse Écologique*. Paris: Doin, 296 pp.

Fedoroff, N. (1966): Contribution à la connaissance de la pédogenèse quaternaire dans le S-O du bassin parisien. *Bull. Assoc. Fr. Etude Quatern.* **3:** 94-105.

Franz, H. (1956): Drei klimabedingte Ranker-subtypen Europas. *Trans. 6th Int. Congr. Soil Sci., Paris* **E:** 135-141.

Franzmeier, D.P., E.P. Whiteside, and M.M. Mortland (1963): A chronosequence of Podzols in northern Michigan. III. Mineralogy, micromorphology, and net changes occurring during soil formation. *Mich. Agric. Exp. Sta. Quat. Bull.* **46:** 37-57.

Gachon, L. (1963): Contribution à l'étude du quaternaire récent dans la Grande Limagne: morphogenèse et pédogenèse. *Thèse, Clermont-Ferrand*, 198 pp.

Gerasimov, I.P. (1956): Les sols des régions méditerranéennes de l'Afrique (du tell). *Trans. 6th Int. Congr. Soil Sci., Paris* **E:** 189-193.

Gerasimov, I.P. (1974a): [Utilization of the concepts of elementary soil processes for a genetic diagnosis of soils.] (in Russian). *Trans. 10th Int. Congr. Soil Sci., Moscow* **6:** 482-489.

Gerasimov, I.P. (1974b): The age of recent soils. *Geoderma* **12:** 17-25.

Gilot, J.C., and Y. Dommergues (1967): Note sur le lithosol calcaire à mor de la station subalpine de la R.C.P. 40. *Rev. Ecol. Biol. Sol* **4:** 357-383.

Glazovskaya, M.A. (1974): [Factors determining profile differentiation in loamy Dernovo-Podzolic soils.] (in Russian). *Trans. 10th Int. Congr. Soil Sci., Moscow* **6:** 102-110.

Guillet, B. (1972): Relations entre l'histoire de la végétation et la podzolisation dans les Vosges. *Thèse Doctorat d'Etat, Univ. Nancy*, 112 pp.

Guillet, B., J. Rouiller, and B. Souchier (1975): Podzolization and clay migration in Spodosols of eastern France. *Geoderma* **14**: 223-245.

Guitian Ojea, F., and T. Carballas (1968): Suelos de la zona humeda española: III. Ranker atlantico. *An. Edafol. Agrobiol.* **27**: 57-73.

Hervieu, J. (1970): Le quaternaire du Nord-Cameroun. Schéma d'évolution géomorphologique et relations avec la pédogenèse. *Cah. O.R.S.T.O.M. Ser. Pedol.* **8**: 295-320.

Hetier, J.M. (1973): Caractères et répartition des sols volcaniques du Massif Central: II. Comparaison Cantal-Chaîne des Puys. *Bull. Assoc. Fr. Etude Sol* (2): 97-109.

Hetier, J.M., M. Rodrigues-Lapa, and F. Le Tacon (1972): Etude micro-morphologique de quelques sols de l'est de la France. *Bull. Assoc. Fr. Etude Sol* (1-2): 49-61.

Icole, M. (1973): Géochimie des altérations dans les nappes d'alluvions du piémont occidental nord pyrénéen. Essai de paléopédologie quaternaire. *Thèse Doctorat d'Etat, Univ. Paris VI*, 328 pp.

Jacquin, F., C. Juste, and P. Dureau (1965): Contribution à l'étude de la matière organique des sols sableux des landes de Gascogne. *C. R. Acad. Agric. Fr.* **51**: 1190-1197.

Jamagne, M. (1963): Contribution à l'étude des sols du Congo oriental. *Pedologie* **13**: 271-414.

Jamagne, M. (1973): Contribution à l'étude pédogénétique des formations loess-iques du Nord de la France. *Thèse de Doctorat, Faculté des sciences agronomiques de l'Etat, Gembloux, Belgium*, 445 pp.

Kovda, V.A. (1965): Alkaline soda-saline soils. In Symp. Sodic Soils, Budapest. *Agrokem. Talajtan Suppl.* **14**: 15-82.

Kovda, V.A. (1973): [*The Principles of Pedology.*] (in Russian) Akad. Nauk. SSSR. Moscow: Nauka, Vol. 1, 447 pp.; Vol. 2, 468 pp.

Kovda, V.A., and G.V. Dobrovolsky (1974): Soviet pedology to the 10th International Congress of Soil Science (the centenary of Soil Science in Russia). *Geoderma* **12**: 1-16.

Kubiena, W.L. (1953): *The Soils of Europe, Illustrated Diagnosis and Systematics.* London: Thomas Murby, 317 pp.

Lamouroux, M. (1971): Etude de sols formés sur roches carbonatées. Pédogenèse fersiallitique au Liban. *Thèse Doctorat d'Etat, Strasbourg*, 314 pp.

Lamouroux, M. (1972): Essai de structuration pour une classification des sols et milieux de pédogenèse. *Cah. O.R.S.T.O.M. Ser. Pedol.* **10**: 243-250.

Latham, M. (1970): Rôle du sol dans la répartition de la végétation au contact forêt-savane dans la région de Séguéla-Vavoua (Côte d'Ivoire). *Thèse D.E.S., Univ. Dijon.*

Le Tacon, F. (1966): Contribution à l'étude des sols d'un massif forestier des basses Vosges. *Thèse Doctorat, Univ. Nancy.*

Maignien, R. (1958): Contribution à l'étude du cuirassement des sols en Guinée française. *Thèse Doctorat d'Etat, Strasbourg*, 311 pp.

Maignien, R. (1959): Les sols subarides au Sénégal. *Agron. Trop.* **14**: 535-571.

Mavrocordat, G. (1971): *Die Böden Rumaniens.* Berlin: Dunker and Humboldt, 156 pp.

Menut, G. (1974): Recherches écologiques sur l'évolution de la matière organique des sols tourbeux. *Thèse Spécialité, Univ. Nancy*, 124 pp.

Nguyen Kha (1973): Recherches sur l'évolution des sols à texture argileuse en conditions tempérées et tropicales. *Thèse Doctorat d'Etat, Univ. Nancy*, 157 pp.

Nguyen Kha, and P. Duchaufour (1969): Note sur l'état du fer dans les sols hydromorphes. *Sci. Sol* (1): 97-110.

Pallmann, H. (1948): Pédologie et phytosociologie. *C. R. Conf. Int. Pedol., Montpellier* (May 1947), pp. 3-36.

Paquet, H. (1969): Evolution géochimique des minéraux argileux dans les altérations et les sols des climats méditerranéens à saisons contrastées. *Thèse Doctorat d'Etat, Univ. Strasbourg,* 348 pp.

Perraud, A. (1971): Etude de la matière organique des sols forestiers de la Côte-d'Ivoire (relations sols-végétation-climat). *Thèse Doctorat d'Etat, Nancy,* 86 pp.

Porta, J. (1973): Mise en valeur de sols salins blancs (Gypsiorthids). Premiers résultats des expériences sur sols des rives de la Giguela (Ciudad Real, Espagne). *An. Inst. Nac. Invest. Agrar. Ser. Gen.* (2): 171-184.

Quantin, P. (1974): Hypothèses sur la genèse des andosols en climat tropical: évolution de la pédogenèse initiale en milieu bien drainé sur roches volcaniques. *Cah. O.R.S.T.O.M. Ser. Pedol.* 12: 3-12.

Richard, J.L. (1961): *Les Forêts Acidiphiles du Jura.* Thesis. Neuchâtel, Switz.: Hans Huber, 164 pp.

Robin, A.M. (1968): Contribution à l'étude des processus de podzolisation sous forêt de feuillus. *Thèse Spécialité, Univ. Paris,* 88 pp.

Robin, A.M., and F. De Coninck (1975): Interprétation génétique d'un horizon pédologique profond ferro-argillique en forêt de Fontainebleau. *Sci. Sol* (3): 213-228.

Ruellan, A. (1970): Contribution à la connaissance des sols des régions méditerranéennes: les sols à profil calcaire différencié des plaines de la basse Moulouya. *Thèse Doctorat d'Etat, Strasbourg,* 482 pp.

Schlichting, E. (1965): Die Raseneisenbildung in der nord-west-deutschen Podzol-Gley-Landschaft. *Chem. Erde* 24: 11-26.

Schröder, D. (1973): Zur Stellung der Gleye und Pseudogleye in verschiedenen Klassifizierungssystemen. In *Pseudogley and Gley, Trans. Comm. V and VI, Int. Soc. Soil Sci., Stuttgart (1971),* pp. 413-419.

Schwertmann, U. (1966): Inhibitory effect of soil organic matter on the crystallization of amorphous ferric hydroxide. *Nature* 212: 645-646.

Schwertmann, U. (1968): Typische Böden in Raum Berlin: I. Die Parabraunerde (Lessivé) aus Würm-Geschiebemergel. *Sitzungsber. Ges. Naturforsch. Berlin* 8: 16-28.

Schwertmann, U., W.R. Fischer, and R.M. Taylor (1974): New aspects of iron oxide formation in soils. *Trans. 10th Int. Congr. Soil Sci., Moscow* 6: 237-249.

Segalen, P. (1973): L'aluminium dans les sols. *Mem. O.R.S.T.O.M.* No. 22, 281 pp.

Servant, J. (1970): Etude expérimentale de l'influence des conditions salines sur la perméabilité des sols: Conséquences pédologiques. *Sci. Sol* (2): 87-101.

Solar, F. (1964): Zur Kenntuis der Böden auf dem Raxplateau. *Mitt. Osterr. Bodenk. Ges. Vienna* 8: 1-70.

Souchier, B. (1971): Evolution des sols sur roches cristallines à l'étage montagnard (Vosges). *Thèse Doctorat d'Etat, Univ. Nancy,* 134 pp.

Sys, C. (1967): The concept of ferrallitic and fersiallitic soils in Central Africa: Their classification and their correlation with the 7th Approximation. *Pedologie* 17: 284-325.

Szabolcs, I. (1969): The influence of sodium carbonate on soil forming processes and on soil properties. In Symp. Reclam. Sodic Soda-Saline Soils, Yerevan, Israel. *Agrokem. Talajtan Suppl.* 18: 37-68.

Targulian, V.O., N.A. Karavayeva, and I.A. Sokolov (1974): [The factors and mechanisms responsible for profile differentiation in soils of boreal regions.] (in Russian). *Trans. 10th Int. Congr. Soil Sci.*, Moscow **6**: 93-101.

Toutain, F. (1974): Etude écologique de l'humification dans les hêtraies acidiphiles. *Thèse Doctorat d'Etat, Univ. Nancy*, 114 pp.

Turenne, J.F. (1970): Influence de la saison des pluies sur la dynamique des acides humiques dans des profils ferrallitiques et podzoliques sous savanes de Guyane française. *Cah. O.R.S.T.O.M. Ser. Pedol.* **8**: 419-449.

Van Wambeke, A. (1971): Hydromorphie et formation de plinthite dans les sols des plaines herbeuses de Colombie. *Mitt. Dtsch. Bodenk. Ges.* **12**: 100-104.

Vieillefon, J. (1974): Contribution à l'étude de la pédogenèse dans le domaine fluvio-marin en climat tropical d'Afrique de l'Ouest. *Thèse Doctorat d'Etat, Paris VI*, 361 pp.

Warembourg, F.R., P. Lossaint, and P. Bottner (1973): L'évolution des sols dans une séquence bioclimatique méditerranéenne-montagnarde sur roche-mère siliceuse: Massif du Mont Aigoual. *Bull. Assoc. Fr. Etude Sol* (1): 49-62.

Zaidelman, F.R. (1974): [*Podzolization and Gleization.*] (in Russian). Moscow: Nauka, 208 pp.

Zonn, S.V., and L.O. Karpachevskii (1964): *Advances in the Theory of Podzolization and Solodization.* USSR Acad. Sci., pp. 1-42.

ADDITIONAL BIBLIOGRAPHY

Aussenac, G., and M. Becker (1968): *Ann. Sci. For. (Paris)* **25**(4): 291-332.

Brethes, A. (1973): Mode d'altération et différenciation pédogénétique sur leucogranite du massif du Morvan. Comparaison avec le Massif vosgien. *Thèse Spécialité, Univ. Nancy.*

Duchaufour, P. (1968): *L'Evolution des Sols.* Paris: Masson, 93 pp.

Duchaufour, P. (1972): *Processus de formation des sols. Biochimie et géochimie.* C.R.D.P. Nancy, 182 pp.

Duchaufour, P., and C. Bartoli. (1966): *Sci. Sol* (2):29-40.

Franz, H. (1960): *Feldbödenkunde.* Vienna: Fromme.

Hetier, J.M. (1971): *Sci. Sol* (2): 51-83.

Jamagne, M. (1964): Introduction à une étude pédologique dans la partie nord du bassin de Paris. *Pedologie* **14**: 228-342.

Kauricheva, I., and I. Gromiko (1974): *Soil Atlas of the U.S.S.R.* Moscow, 164 pp.

Kovda, V.A., and E.M. Samoilova (1969): Some problems of soda salinity. *In*: Symp. Reclam. Sodic Soda-Saline Soils, Yerevan, Israel. *Agrokem. Talajtan Suppl.* **18**: 21-36.

Kundler, P. (1965): *Waldbödentypen.* Hanover: Neumann Verlag, 180 pp.

Leneuf, N. (1959): L'altération des granites calco-alcalins et des granodiorites en Côte-d'Ivoire forestière et les sols qui en sont dérivés. *Thèse Doctorat d'Etat, Univ. Paris*, 210 pp.

Muckenhausen, E. (1962): *Entstehung, Eigenschaften und Systematik der Böden der Bundesrepublik Deutschland.* Frankfurt-Main: DLG Verlags, 148 pp.

Pelisek, J. (1961): *Atlas hlavnich pudnich typu.* C.S.S.R., Prague, 440 pp.

Quantin, P. (1972): *Cah. O.R.S.T.O.M. Ser. Pedol.* **10**: 123-151.

Scheffer, F., and P. Schachtschabel (1966): *Bödenkunde* (6th ed.). Stuttgart: Enke, 473 pp.

Servant, J., and E. Servat (1971): *Note, Int. Soc. Soil Sci., Sevilla*, 16 pp.

Uggla, H. (1965): *Gleboznawstwo lesne Szczegolowe.* Warsaw, 400 pp.

Vedy, J.C. (1973): Relations entre le cycle biogéochimique des cations et l'humification en milieu acide. *Thèse Doctorat d'Etat, Univ. Nancy*, 116 pp.
Warembourg, F. (1969): Sur la dynamique des sols dans les Cévennes méridionales siliceuses. Une séquence altitudinale dans le massif de l'Aigoual. *Thèse Spécialité, Univ. Montpellier*, 142 pp.

SOIL MAPS

Soil Map of France (1/100,000) (Director: M. Jamagne, I.N.R.A., Versailles).
Saint-Dié Sheet (Centre de pédologie biologique, C.N.R.S., Nancy, 1973).
Nancy Sheet (Centre de pédologie biologique, C.N.R.S., Nancy).
Dijon Sheet (I.N.R.A., Dijon, 1973).
Vichy Sheet (Service de cartographie des sols, Montpellier: F. Servat).
Perpignan Sheet (Service de cartographie des sols, Montpellier).
Laon Sheet (Station agronomique de l'Aisne, Laon: M. Jamagne and J. Hébert, 1973).
Nîmes and Montpellier Sheets (Compagnie nationale du bas Rhône-Languedoc, Nîmes, 1974).
Map of Massif de Haye (1/25,000) (C.N.R.S., Nancy: M. Gury, 1972).
Soil Map of the World (F.A.O.-UNESCO: R. Dudal).
Soil Map of the World (U.S.S.R.: I.P. Gerasimov).
Soil Map of Poland (Warsaw, 1974).
Soil Map of the World-Europe (Rome: F.A.O.-UNESCO, 1967).
Soil Map of the World-South America (Paris: F.A.O.-UNESCO, 1971).
Atlas zur Bödenkunde (Ganssen, R., and F. Hädrick, Hochschulatlanten, Bibliographisches Institut, 301a-301e).

SOIL TAXONOMIES

Soil Classification in the Soil Survey of England [Avery, B.W. (1973): *J. Soil Sci.* 24: 324-338].
Classification française des sols (C.P.C.S., 1967).
The Soil Classification System of the Federal Republic of Germany [Mückenhausen, E. (1965): *Pedologie* No. Spec. 3: 57-74].
Soil Taxonomy (U.S. Dept. of Agriculture, Soil Conservation Service, Agric. Hbk. 436, Washington, D.C. 1975).
Legend F.A.O.-UNESCO Soil Map of the World (Rome: F.A.O., 1973).

INTERNATIONAL CONFERENCES AND CONGRESSES

(Scientific Excursions)

6th International Congress of Soil Science, Paris, France, 1956.
7th International Congress of Soil Science, Madison, Wisconsin, U.S.A., 1960.
8th International Congress of Soil Science, Bucharest, Romania, 1964.
10th International Congress of Soil Science, Moscow, U.S.S.R., 1974.
National Soil Survey Conference, Romania, 1969.
Conference on Mediterranean Soils, Madrid, 1966.
Conference on Hydromorphic Soils, Stuttgart, 1971.

NOTE: This bibliography is complementary to that of the author's previously published books, in particular to that of Duchaufour (1970).

INDEX

175